Elixir

Also by Brian Fagan

Elixir

A History of Water and Humankind

Brian Fagan

BLOOMSBURY PRESS

New York   Berlin   London   Sydney

Published by Bloomsbury Press, New York

All papers used by Bloomsbury Press are natural, recyclable products made from
wood grown in well-managed forests. The manufacturing processes conform to
the environmental regulations of the country of origin.

LIBRARY OF CONGRESS CATALOGING-IN-PUBLICATION DATA
Fagan, Brian M.
Elixir : a history of water and humankind / Brian Fagan.—1st U.S. ed.
p.   cm.
Human history of water
Includes bibliographical references and index.
ISBN-13: 978-1-60819-003-4 (alk. paper)
ISBN-10: 1-60819-003-X (alk. paper)
1. Water. 2. Water—History. 3. Water—Social aspects—History.
4. Water and civilization—History. I. Title. II. Title: Human history of water.
GB671.F34 2011
553.7—dc22
2010032082

First U.S. edition 2011

3   5   7   9   10   8   6   4   2

Typeset by Westchester Book Group
Printed in the U.S.A. by Quad/Graphics, Fairfield, Pennsylvania

For

My best girls . . .

Alexa, Ana, Juno, Lesley, and Pipette

"Now John," quod Nicholas, "I wol nat lye;
I have yfounde in myn astrologye,
As I have looked in the moone bright,
That now a Monday next, at quarter nyght,
Shal falle a reyn, and that so wilde and wood
That half so greet was nevere Noes flood.
This world," he seyde, "in lasse than an hour
Shal al be dreynt, so hidous is the shour.
Thus shal mankynde drenche, and lese hir lyf."

 —Geoffrey Chaucer, "The Miller's Tale" (c. 1390 C.E.)[1]

The tending of the rice plants, from the time of replant-
ing to the harvest, the sensuality of wading in the warm
mud, the concentration of nurturing the fragile plants
as one would a child, the sense of continuity one gets
from finding under one's fingers, during the weeding
of the new crop, the half decayed vestiges of the last
crop buried in the terrace to fertilize the soil, all these
experiences . . . are experiences of the senses and of
the body.

 —Arlette Ottino[2]

Contents

Part V   Gravity and Beyond

Preface

MANY YEARS AGO, three San hunters and I trekked across southern Africa's arid Kalahari Desert on a searingly hot morning at the end of the dry season. We had stalked duiker since dawn, searching unsuccessfully for the elusive antelope in the shady thickets where they settled as the sun climbed in the sky. My companions moved effortlessly, apparently without fatigue or thirst, as I paused to take regular swigs from my water bottle. We came to a dry watercourse and a solitary grove of trees that cast the only shade for miles around in the seemingly waterless landscape. The men paused to rest. One of them examined the dry streambed and dug into the sand with his wooden digging stick. At first, the soil was dry, then it was damp, then, miraculously, water appeared. The hunter crouched, swept the precious liquid up with his hands, and drank deeply. His companions followed; so did I, allowing the water to flow over my sweating face and hands. I have never felt such a close, sensuous connection with the most vital elixir of life. My companions had found water where I had thought there was none. As I got to know them better, I learned something of a new way of looking at the landscape: as an edible and drinkable persona, rich in liquid-bearing plants and hidden water. I realized that San existence depended on the distribution of water across the landscape and on the ancient traditions that passed water knowledge from one generation to the next. Since that defining moment, water has always had a profound significance to me.

I CLOSED MY eyes, listened, and was calmed. The gentle riffle of fountains, of softly flowing water, permeated my senses on that hot afternoon in Granada's Alhambra Palace. I'd walked through the hilltop park, famous for its nightingales, never far from the cooling sounds of running water. A few minutes later, I stood in the Court of the Pond, where a great rectangular pool lined by myrtles cools the palace. Goldfish swim in the calm waters. When the crowds are gone, the courtyard exudes a profound serenity. Looking out over the city and the sunburnt hills in the distance, I marveled at the lush oasis around me. Later, I learned that the Islamic architects had built a five-mile (eight-kilometer) conduit to bring water from the Darro River.

The marvels continued at Jannat al-'Arif, the Generalife Palace, across a nearby ravine. Nasrid sultan Muhammad III built his summer palace during the first decade of the fourteenth century. He spared no expense on the magnificent landscaping, one of the oldest surviving Islamic gardens in the world. In the Water-Garden Courtyard, arcing jets of water play on your senses as they soar, sparkle, and tumble into a long, rectangular pond. This is a place of colonnades and pavilions for leisure and contemplation; flower beds press on the shimmering pool; the gentle sounds of flowing water add to the impression of paradise. And a human paradise it is, in a harsh land of steep terrain and unpredictable rainfall.

The four rivers of the Islamic paradise are "rivers of water incorruptible," which nourish "gardens beneath whose trees rivers flow."[1] The Prophet greets the faithful near a pool in paradise named Kauthar. Granada's resplendent gardens are a green oasis, an ideal of serenity and well-being. They epitomize the close relationship between humans and water, common to all societies on earth, yet expressed in all manner of ways, some of them startlingly direct and intimate.

Water: It caresses and comforts us, provides sustenance and refreshment, is something that humanity has cherished since the beginning of history, and means something different to everyone. Water gives us pleasure, as it does at the Alhambra, and has profound sacred qualities. It figures largely in many holy and special places—the soft murmuring of a sacred spring at Delphi, in Greece; the reflecting pools of India's Taj Mahal; the reservoirs that surround Angkor Wat, in Cambodia, symbolizing the primordial waters of the universe; the font for holy water in Christian cathedrals. Water evokes serenity, harmony and peaceful existence, the very essence of life, and is commemorated by grand shrines and elaborate rituals in honor of the deities that ensure the continuity of water—and life itself.

Water: We turn a faucet, and it is there for drinking, something we take completely for granted. So commonplace is water in our daily lives that we are indifferent to it and have been for a long time. Years ago, Rachel Carson wrote that "in an age when man has forgotten his origins and is blind even to his most essential needs for survival, water along with other resources has become the victim of his indifference."[2] Of all the resources that we rely on for survival in today's world, water is the least appreciated and certainly the most misunderstood. For generations, we in the industrial West have just assumed that fresh drinking water is ours to enjoy and to use with dazzling promiscuity in any way we wish. This shouldn't surprise us in an urban age when almost everyone buys their food as packaged commodities from supermarkets, in an era when many city children never see a cow except in pictures. Water is like beef, milk, and pasta, an integral part of our lives that we never think about—a great mistake.

History tells us that a defining moment in our use of the earth's water came with the development of steam power and turbine pumps and the harnessing of fossil fuels during the Industrial Revolution of the late eighteenth and nineteenth centuries. For the first time, rapidly expanding and industrializing nations had access to enormous water supplies deep underground, not only for domestic use but also, and more important, for agriculture and industry. Much of this water mining took hold in arid and semiarid lands, where lakes, rivers, and springs could never support intensive agriculture or large cities. An orgy of consumption ensued and continues to this day, amid alarming signs in areas like the American West that the supply of underground water is finite and not being replaced at anything like a sustainable rate. In a warming world where we know that higher temperatures historically tend to be associated with prolonged droughts, the alarm flags of pending water shortages flap in the environmental wind. Yet most of us living in drier environments still take water for granted and are in a state of denial about the impending crisis. California is a case in point: It is suffering from multiyear droughts and greatly diminished water supplies, yet its farmers demand full allocations of irrigation water at heavily subsidized prices. Even in good rainfall years, water for all our needs is in short supply. There are calls for desalinization plants to supplement nature's supplies in arid environments, as well as for additional dams and more long-distance aqueducts. All of this misses the point. The distribution of the world's water does not match the areas of greatest need, many of which lie in arid and semiarid lands.

A huge academic and popular literature surrounds water and the impending crisis, complete with both general and more specific recommendations. There's been much wringing of hands, but it is only now that water conservation is moving to center stage. It is becoming clear that all of us, whether farmers, governments, manufacturers, or common folk, need to do with less in the future, at a far more profound level than sweeping instead of washing down our driveways and watering our lawns less frequently. And it may be here that we can learn from our forebears, for they lived in worlds where water was often scarce, frequently hard to obtain, and treated with great respect. They lived long before the days of pumps and artesian wells, but they knew everything there was to know about making do with finite water supplies and about the force that propels water: gravity. The water managers of the past, whether village farmers, well-connected officials, or court engineers, knew that water is a vital and pitiless force in human life. We can consume, but can never completely tame, it. As long ago as the sixth century B.C.E., the Taoist philosopher Lao-tzu wrote in his *Tao Te Ching*, "Nothing in the world is as soft and yielding as water. Yet for dissolving the hard and inflexible, nothing can surpass it. The soft overcomes the hard; the gentle overcomes the rigid."[3] History teaches us again and again how right he was.

This is a book about human relationships with water in the past, as far as I can tell the first such work to tackle the subject on a global scale. The idea for such a project revolved in my mind for decades, a seed germinated unconsciously many years ago when I was living among subsistence farmers in central Africa a few miles downstream of a major hydroelectric dam. More seeds of the idea took root in later years. While working in Africa, I experienced water deprivation firsthand on many occasions. Walking long miles on archaeological surveys with only a water bottle on my hip when the temperature was in the eighties taught me hard lessons about dehydration. I grew my beard—now forty-four years old—after several water-short weeks during fieldwork in Uganda when washing was a real luxury. Years later, crossing the Atlantic in a sailboat was a memorable experience, especially when one of our spare water cans sprang a leak in mid-ocean, a nerve-racking incident by any standards. More recently, I've done a great deal of bicycling on hot California days. Such trips have reminded me forcibly about the need to keep hydrated. The idea finally took root when my wife, a master gardener, relandscaped our yard with drought-resistant plants and pulled out all our lawns. Literally the day she completed the project, I heard a radio story about plans for a water park along the drought-challenged coastline south of Los Angeles, and another one about a wealthy resident of our city who spent thirty thousand dollars on tanked-in water for his huge lawns. These tales of wasteful consumption tipped the scale. For some years, I had been writing a series of books on ancient climate change, especially on the Medieval Warm Period and on El Niños, which brought me up against the issue of prolonged droughts time and time again. With my hard-won experiences with water, lengthy exposure to the vicissitudes of ancient climate change, and long decades of teaching archaeology on a global scale, it was time to tackle what was to me a virtually unknown subject. The resulting journey has given me a unique perspective on ancient societies of all kinds and the ways in which they tried to achieve the holy grail of any water manager, whether an African villager or a Chinese irrigation engineer: long-term sustainability.

As I RESEARCHED *Elixir*, I was struck by how little most people's relationship with water changed over the thousands of years from the first appearance of agriculture some twelve thousand years ago into medieval times and beyond. Even today, millions of subsistence farmers live from harvest to harvest, from one rainy season to the next, dependent on unpredictable water supplies from the heavens. This led me to think of the history of humans and water in terms of three stages, which overlap with one another. The first goes back to the remote past and endures in places today. Water was an unreliable, often scarce, and always valuable

Table of Major Developments and Events

c. 10,500 B.C.E.	Agriculture in the Near East
Unknown date	Earliest efforts at furrow irrigation
Before 7000 B.C.E.	Earliest known well in Cyprus
Before 5000 B.C.E.	Furrow irrigation in Wadi Faynan, Jordan
	Irrigation in Choga Mami, Iraq
Before 4500 B.C.E.	Irrigation agriculture well established in southern Mesopotamia and Egypt
3800 B.C.E.	Southern Mesopotamia becomes drier
c. 3100 B.C.E.	Sumerian (Mesopotamian) and Egyptian civilization come into being
2700–1900 B.C.E.	Mature Harappan civilization in the Indus Valley
2334 B.C.E.	Sumerian civilization overthrown by King Sargon of Agade
2297–2198 B.C.E.	Alleged dates of Yu the Great, pioneer Chinese hydrological engineer
c. 2000 B.C.E.	Extensive irrigation works by the pharaohs in Egypt's Fayum Depression
c. 1600 B.C.E.	Sophisticated water management at the Minoan Palace of Minos, Knossos, Crete
c. 1500 B.C.E.	Maize agriculture and irrigation in the American Southwest
After 1366 B.C.E.	Assyrians rise to power in northern Mesopotamia
883–859 B.C.E.	Assyrian monarch Assurnasirpal II builds canals, a tunnel, and irrigation works at Kalhu (Nimrud), northern Iraq
714 B.C.E.	The Assyrian ruler Sargon II admires a *qanat* in Urartu, Iran, the first record of such a water device
704–681 B.C.E.	Assyrian king Sennacherib's irrigation works at Nineveh

612 B.C.E.	Nineveh sacked by Babylonian conquerors
Sixth century B.C.E.	Marib dam, Yemen, constructed
594 B.C.E.	Solon promulgates water regulations in Athens
560 B.C.E.	First large-scale Chinese canal digging
After 550 B.C.E.	*Qanats* spread with the Achaemenid Empire (550–330 B.C.E.)
539 B.C.E.	Cyrus the Great captures Babylon
530 B.C.E.	Eupalinos constructs his water tunnel on Samos
323 B.C.E.	Death of Alexander the Great
323 B.C.E.–224 C.E.	Seleucid and Parthian rule in Mesopotamia
312 B.C.E.	Aqua Appia, Rome's first aqueduct, built
Fourth century B.C.E.–eleventh century C.E.	Anuradhapura flourishes in northern Sri Lanka
300 B.C.E.	Ptolemy I Soter founds a library and museum in Alexandria, an important center of technological development
258–237 B.C.E.	Major drainage and irrigation works in Egypt's Fayum under Ptolemy II and III
256 B.C.E.	Construction of the Dujiangyan irrigation system, Sichuan Province, China (still in use)
110 B.C.E.–sixth century C.E.	Himyarite kingdom in Yemen
First century B.C.E.	Vertical waterwheels widespread in the eastern Mediterranean world
25 B.C.E.	Marcus Vitruvius Pollio publishes his *De Architectura*
11 C.E.	Sanyangzhuang inundated in northern China
50	Segovia, Spain's aqueduct constructed
97	Sextus Julius Frontinus appointed water commissioner for Rome
224–651	Sassanian Empire

c. 400	Possible irrigation agriculture in Bali, known to be well established by the eighth century
c. 450	Hohokam farmers in the American Southwest
622	The prophet Muhammad's hegira to Medina (Yathrib)
640	Islam spreads to Syria, Palestine, Egypt, and Libya
711	Islamic conquest of Spain (al-Andalus) begins
1113	Founding of Angkor Wat, Cambodia, by Khmer king Suryavarman II
1300	Overshot water mills in widespread use in medieval Europe
Early fourteenth century	Vijayanagar founded in southern India
1431	Khmer civilization based on Angkor collapses
1492	Al-Andalus becomes part of Christian Spain
c. 1500	Hohokam societies implode
c. sixteenth century	Irrigation in Engaruka, Tanzania. Similar furrow irrigation probably well established elsewhere in sub-Saharan Africa by this time
1696–1771	Life span of Chen Hongmou, influential Chinese irrigation expert
1752	Primitive Newcomen steam engines pump water for London
1753	Frederick the Great completes drainage of the Oder Marshes
c. 1762–64, 1769	Invention of the spinning jenny and water frame for cotton spinning
1776	James Watt develops the first commercially viable steam engine as part of the Industrial Revolution

1907–1924	New York's Catskill Aqueduct system developed
1913	Opening of the Owens Aqueduct, which supplied water to Los Angeles
1931–36	Construction of the Hoover Dam

resource, so precious that it was sacred in almost every human society. The second stage began in part about two thousand years ago, accelerated during the European Middle Ages, and reached fruition during the Industrial Revolution, when water became a mere commodity. Eventually, we developed the technology to pump water from deep beneath the earth and to harvest it on a large scale. Water became something we could exploit without ever worrying about how much we used. Now, during the early twenty-first century, we are in a third stage where we are finally realizing that water is a finite resource, something to be conserved and treated with a respect, even reverence, that we haven't displayed in a long time.

With these stages in mind, *Elixir* revolves around three broad themes. The first is gravity, the fact that water flows downslope, from a higher point to a lower one. There was no other way of moving water except for small-volume pumps and waterwheels until the Industrial Revolution. Even today, gravity plays a central role in water management everywhere, even with long distance aqueducts such as those that feed vast cities like Los Angeles and Phoenix. Ancient Roman and Greek engineers were maestros of gravity-fed water delivery. So were the Chinese and the Inca of Peru. Dozens of smaller-scale societies and village farmers around the world still use gravity to irrigate their fields and to water their beasts. Some of them have maintained sustainable water supplies for centuries and are capable of doing so indefinitely if other users don't hijack their sources with pumps and earthmoving machinery. This book is a history of the triumph of gravity, the silent, ubiquitous force behind nearly all human relationships with water. Gravity lies behind the flexible, inexorable forces of water, but those who take advantage of it don't pretend to control them.

The second theme is the close relationship between ritual and water management of all kinds. Water has a special place in all human societies. It's the essence of fertility and growth, of sustained life, associated with cleansing and renewal, with the spiritual forces of the cosmos. We've worshipped it and celebrated its magical, flowing qualities, commemorated its mystical dimensions. We're in awe of water. As fisherfolk, sailors, and surfers we respect its mysterious attractions. At the same time, it has an indispensable role in human life, for it lies behind everything we do, from cooking food and washing clothes to agriculture, cattle herding—even baseball, tennis, and golf. Water is one of the few cultural universals, inspiring a profound mingling of ritual and day-to-day use.

Ritual and religious beliefs surrounding water permeate every stage of human life from birth to death. Christian infants are baptized in holy water. The ancient Egyptians and the Maya believed that their worlds began in the still waters of the primordium. Bronze Age farmers on the shores of the North Sea three thousand years ago thought that their ancestors dwelled beneath the heaving waters of the ocean—the realm of the dead. Australian Aboriginal existence and much of its ritual revolves around knowledge of where water lies across the landscape. This book goes beyond hydrology and human ingenuity to explore some of the spiritual ties between ancient human societies and water.

The third theme is technology versus sustainability, efforts at living within one's hydrological means. The past teaches us much about water management and has significant lessons for today and the future. Early theories about ancient irrigation conjured up dramatic images of slaves laboring waist deep in mud at the bidding of a harsh supervisor's whip. Such scenarios featured anonymous regiments of laborers, who transformed landscapes and created the underpinnings of preindustrial civilizations like those of Mesopotamia and China. Famed movie producer Cecil B. DeMille might have dreamed of these lingering stereotypes, which are inaccurate at best but occasionally still haunt the history books. In fact, some of the past's most effective irrigation works involved small villages and came online long before any great ruler ascended his throne. Simple furrow irrigation, the diversion of stream and river water into nearby fields, is almost as old as farming itself, a logical

step for villagers grappling with irregular rainfall, potential crop failures, and long dry seasons. Like their remote hunting-and-gathering
ancestors, they were experts at managing risk, at sustaining life by turning to wild plants or hunting, by developing simple ways of bringing
water in gently sloping furrows to their fields. Such self-sustaining water management survived far from the historical limelight in subsistence
farming communities throughout the world for thousands of years. It
still flourishes today in many places where the diesel engine and the turbine pump are but expensive dreams. This type of agriculture probably
reaches its most sophisticated iteration among the rice farmers of Bali,
in Southeast Asia. Here, the allocation and management of water is in
the hands of the farmers themselves and the deities, for ritual and irrigation are completely intertwined.

Contrary to popular belief, village irrigation and local control of
water remained very important long after the appearance of urban civilization in China, Egypt, and Mesopotamia. As I argue, the main interest of rulers and their officials was not in irrigation and water
management as such, but in the food surpluses that they produced,
which supported the state. Taxes collected in grain or labor were the
mainstay of preindustrial civilizations everywhere, where the many
toiled for the benefit of the few. The Egyptian pharaohs presided over
an agrarian kingdom, where everyone had strong ties to ancestral villages and to the land. Each summer, an inundation flooded basins in the
floodplain in a routine that went back centuries before the first divine
king ruled Upper and Lower Egypt. The grain and labor of the villagers
built the pyramids and great temples, and when there was drought in
about 2200 B.C.E., Egypt fell apart for more than a hundred years. In
response, later pharaohs embarked on much-larger-scale irrigation
works in places like the Fayum Depression, to the west of the Nile, and
invested heavily in granaries. The same was true in Mesopotamia, where
the early cities were little more than agglomerations of villages that
came together in the face of increasingly scarce floodwaters.

Circumstances changed as climate changes in the Indian Ocean
caused major shifts in monsoon patterns after 2800 B.C.E. Summer no
longer brought welcome rain to Mesopotamian farmers. Every village
now relied more heavily on spring floods for irrigation water. More

centralized control of irrigation works and water became imperative to ensure equitable distribution of water, especially in increasingly frequent drought years. In both Egypt and Mesopotamia, an army of officials now supervised canal maintenance and harvests, for careful water management made an immediate difference between famine and plenty for increasingly dense rural and urban populations. Inevitably, the scale of irrigation works grew and grew, with the Assyrians and their small armies of prisoners of war and with the Sassanians after them, who, like some Chinese emperors, transformed entire landscapes, sometimes with devastating ecological consequences. Southern Mesopotamian farmers had confronted increasing salinization from the early days of irrigation, handling it with, among other methods, systematic fallowing. But Mesopotamia later became a salt-ridden wilderness after the Sassanians took up five times more land under more irrigation than ever before in an insatiable quest for tax revenue and wealth. The result was promiscuous water usage and the ecological devastation of thousands of acres. The analogies with today's unthinking water usage in the American West are sobering. Some of China's large-scale irrigation works along the Huang River have contributed to the ecological crisis of endemic silting and water shortages that inflict the region today. Viewed in retrospect, such failed projects offer a sobering example of what happens to cities, farmers, landscapes, and indeed entire societies when sustainability evaporates.

There were notable triumphs of technology and sustainability, too, as in ancient Greek cities that made use of water-rich karst formations to provide reliable, perennial water supplies for domestic use, agriculture, and waste disposal. The Romans, with their aqueducts, took a long-established technology and turned it into an architectural art form, as much as a matter of prestige and for public bathing as for reliable water supplies. Many of their water systems were sustainable over many centuries, even if they were wasteful. Roman expertise with waterwheels and mills formed an important legacy for medieval Europe. The Romans did not invent such devices, but they used them with near-industrial effect grinding grain for Rome and communities like Arles.

Anyone relying on gravity-derived water supplies had no option but to maintain a high degree of self-sustainability, especially in arid lands.

The groundwater-tapping *qanats* of Iran and the *puquios* of the southern Andes—gravity-fed, humanly dug tunnels that tapped groundwater— were brilliant, simple inventions that determined where cities, towns, and villages were to lie. With such small-scale devices, you brought the settlement to the water. The Assyrians, Greeks, and Romans took another tack, sometimes using aqueducts for the reverse purpose. The peasants and engineers of medieval Islam were perhaps the most ingenious of all preindustrial water managers. They used the simplest of technologies to wrest water from seemingly arid landscapes. As has become apparent in recent years, almost all water management in the Islamic world was ultimately based in the village, in the small farming community that designed its own water system based on the local contours, using dams and canals, large and small. When it came to cities, mosques, and palaces, the same simple, gravity-based technologies, combined with waterwheels and the labor of human and beast, came into play.

By the sixteenth century, the European and Mediterranean worlds had reached the limits of gravity-driven water systems. Two centuries later the Industrial Revolution changed the entire water equation for humanity with pumps and earthmoving machinery that opened up hitherto-inaccessible water supplies in the depths of the earth. In China, an eighteenth-century Qing Dynasty official named Chen Hongmou paid thousands of laborers to drain entire landscapes and dig canals that brought mountain water to lowland fields. Unbeknownst to him, Chinese water management was close to a point that it reached some time later, when nothing further could be done to manage water with the existing technology. Only a century after Chen's hard work, a new technological era had dawned on the other side of the world, which allowed governments to reshape entire landscapes and transport water over distances and mountain ranges as never before, with only a fraction of the unskilled human labor used by Chen.

There's another major facet of humans' relationship with water that I don't explore in these pages: the use of water for both transport and voyaging, be it over a lake, up a river, or across an ocean. Until the domestication of the camel in the first millennium B.C.E., waterways, especially major rivers like the Euphrates, the Tigris, the Nile, and the

Yangtze, provided much better ways of traveling than on foot or on the back of an ass or horse that required regular watering. Great rivers became highways from one village or city to another and for moving goods of all kinds over distances long and short. These journeys spawned breakthrough inventions like the planked boat and the sail. By 3000 B.C.E., sailing vessels were carrying timber and other heavy cargoes long distances with the aid of prevailing winds and river currents. From these inland ventures, it was a small step to coasting along Mediterranean and Persian Gulf shores, and from there to undertaking voyages to lands, known and unknown, invisible over the far horizon.

In fact, seafaring in a systematic way began much earlier than civilization—at least fifty-five thousand years ago off the coast of Southeast Asia. It took many millennia for human societies of all kinds to decipher the oceans. How exactly they did so remains one of the fascinating conundrums of history. What caused people to cross open water to uninhabited land dimly visible offshore on clear days? Was it food shortages or a lack of agricultural land? Or was it simply human curiosity? How and why did people in the southwestern Pacific decide to sail in outrigger canoes in search of islands deep in the unknown Pacific three thousand years ago? What were the beliefs and spiritual relationships that sustained human societies as they colonized the Aleutian Islands or the Aegean Sea, in the Mediterranean? I'll explore these numerous questions in my next book, which will be, in many respects, a sequel to this one.

ELIXIR EXPLORES ALL manner of human societies, well known and obscure. We cannot understand the complex relationships between humans and water without traveling far beyond the classic archaeological and historical stomping grounds of Europe, the Mediterranean, and Central America. Nor can we, as I once thought we could, visit the history of humans and water along a linear chronological track. My own experiences in Africa and elsewhere, and even a superficial examination of the literature on ancient water, soon made it clear that some extremely simple water-management approaches, such as furrow irrigation, not only nourished farmland many thousands of years ago but also thrive in self-sustaining societies into the twenty-first century. For this reason, the early chapters of the book examine furrow irrigation in a few ancient and still-existing subsistence-farming societies. The latter are self-sustaining in a world where much water management has moved far beyond sustainability. Some of these societies, like the Pokot of Kenya, administer their water systems by consensus and discussion. Others, like the rice farmers of Bali, depend on ancient rituals and long-established administrative and religious mechanisms to share water from upslope with farmers living much further downstream. The Bali system is so effective that Dutch colonial authorities and their successors failed to come up with anything more efficient. Then there's the remarkable case of the Hohokam of Arizona's Sonoran Desert, who flourished effortlessly in one of the driest environments in the Americas for a thousand years before prolonged droughts during the Medieval Warm Period caused a now-much-more-elaborate farming society to implode. The contrast with the vast urban sprawl of today's Phoenix, which lies atop the Hohokam's ancient irrigation works, is both disturbing and enlightening.

The middle chapters of the book do form a chronological gradient, telling the complex story of water management in the Mediterranean world. Here, we navigate some relatively familiar historical territory, and also examine much that has rarely emerged from the specialist literature. Like the study of ancient climate, archaeological studies of broad landscapes, as opposed to individual cities, towns, and villages, have gone through a scientific revolution in recent years. Today's archaeologists wear out shoe leather like their predecessors, but they now have a far wider range of tools to draw on. Aerial photographs, satellite maps, global positioning systems (GPSs), and other tools help them locate long-vanished canals and house mounds that give more complete pictures of ancient landscapes. When these are combined with historical data such as clay tablets, inscriptions, and papyri, it's astounding how much information can be teased from ancient irrigated landscapes, although many of them have, of course, been destroyed by later intensive farming activity. There were surprises for me here, too. For instance, startlingly few changes took hold in Egyptian water management over millennia; Sumerian irrigation in Mesopotamia was much less ambitious than I had assumed. The hydrological expertise of the Minoans of Crete, and of the ancient Greeks, was an eye-opener, for the Romans copied many of their water devices, including siphon technology, which had originated much earlier in the Near East on a more modest scale.

From the Mediterranean world, we travel to India and Southeast Asia, to a world of cisterns and dams, where monsoon rains played a central role in maintaining sustainability. Here again, new discoveries are rewriting history, as the scale of water-management works at sites like Anuradhapura, in Sri Lanka, and the city of Vijayanagar, one of the largest cities in ancient South Asia, is becoming apparent after centuries of relative historical obscurity. Then there's Angkor, in Cambodia, where water is seemingly abundant, but where recent fieldwork in the hinterland surrounding Angkor Wat shows that monsoon failures and drought drastically affected the Khmer Empire and may even have contributed to its collapse. China offers another dramatic contrast, a land of two worlds: the south, with its abundant water and rice agriculture, and the far more challenged north, where famine, drought, and water shortages have haunted village farmers since the beginnings of farming life. The ambitious plans that today's China has for moving water from the south to the north have deep roots in history, where emperors' minions set thousands to work building long canals and miles of dikes.

Water management in ancient America, discussed next, offers striking parallels between the water problems at Angkor and those of the ancient Maya, to whom irrigation was unknown and whose farmers relied on raised fields in swamps and on tropical subsistence farming. In the end, prolonged droughts were one of the causes of the collapse of much of Maya civilization in the tenth century C.E., whereas the Andeans survived drought after drought along the arid Peruvian coast through conservative, careful water management. The Inca, high in the Andes, were water engineers of genius and triumphant users of gravity.

Finally, we return to the Near East and West. Humans have always lived in unpredictable environments, where water resources lie irregularly distributed across the landscape. That landscape can be arid, with only seasonal rainfall or virtually no rainfall at all. Such was the world of Islam, whose water engineers designed gardens that truly offer a blueprint for paradise in a water-deprived world. However, Islamic water management faltered, in part because of drier conditions and political upheavals, as well as growing populations, but primarily because their engineers came up against the limitations of their technology and lacked the circumstances to innovate. Instead, it was medieval Europe, with its plentiful water, that ultimately developed the technologies that changed human relationships with water during the nineteenth century.

THIS BOOK IS about changing human relationships with water over thousands of years. Our story is a complex meld of climate change, gravity, human modifications of the natural environment, and technological innovation, kept in balance by intricate ritual observance and religious belief. There are many smaller-scale societies around the world that manage their water in sustainable ways, and will continue to do so if the greedy maw of industrial civilization does not shrink groundwater levels and divert streams. Even during drought cycles, the most resilient of these societies survive. As the Hohokam teach us, it was ultimately their success, reflected in rising populations and increased social complexity, that did many of them in. Then there are the rest of us, who take water for granted as it comes out of faucets or stands on supermarket racks in plastic bottles. Most of the water we pump from deep aquifers supports agriculture, helping provide the food that we buy in the same markets. We rarely think of water as a finite resource that may one day dry up as we play a round of golf, fill our swimming pool on a hot summer day, or water our lawn. As small regiments of writers quite rightly lament, we're long past the frontiers of water sustainability, especially in areas like the American West, Australia, and the Near East. We live in the industrial age of water as a commodity, yet alongside us thrive much smaller societies that use water wisely, as they always have. Now we are entering a new era caused by our own wastefulness. The new era, of carefully husbanded water supplies, is one of conservation. History teaches us that the societies that last longest are those that treat water with respect, as an elixir of life, a gift from the gods. We seem to have forgotten this compelling lesson.

Author's Note

Place names are spelled according to the most common usage.

Archaeological sites and historic places are spelled as they appear most commonly in the sources I used to write this book. Some obscure locations are omitted from the maps for clarity; interested readers should consult the specialist literature.

The references and notes tend to emphasize sources with extensive bibliographies, to allow the reader to enter the more specialized literature if they desire. All radiocarbon dates have been calibrated, and the C.E./B.C.E. convention is used.

Every reasonable effort has been made to contact copyright holders for the illustrations. Anyone with questions should contact the author.

PART I

Canals, Furrows, and Rice Paddies

Living from one rainy season to the next: In which we visit Australian Aborigines and the earliest water management of all, explore furrow irrigation in ancient and modern sub-Saharan Africa, trace ancient canal systems in the Arizona desert, and decipher the intricacies of ritual and rice farming in Bali.

CHAPTER 1

The Elixir of Life

INGOMBE ILEDE, "the place where the cow sleeps," lies on the floodplain of central Africa's Middle Zambezi Valley, near the hamlet of Lusitu, in Zambia, downstream of the rugged Kariba Gorge (map: figure 2.1). Four years in the making, the giant Kariba Dam blocked the Middle Zambezi River in 1959, forming a lake that covers 2,150 square miles (5,580 square kilometers) as part of a huge central African hydroelectric project.[1] The government of what was then Northern Rhodesia forcibly resettled fifty-seven thousand Gwembe Tonga farmers, whose ancestors had lived upstream of the dam site for many centuries. Many of them came to villages around Lusitu. They found themselves in a much less productive location, for they had always farmed small riverside gardens watered by the Zambezi's spring floods. Now the floods were gone. The reengineered river flowed between tall banks, its gravity-propelled momentum controlled by giant dam sluices. The only water for the villagers' fields came from five months of irregular rainfall in a low-lying environment where droughts were commonplace.

Almost immediately, there was hunger and thirst. Belatedly, the colonial government sunk boreholes. The groundwater was unpleasantly salty, so the engineers decided to pump water from the river instead. A network of small pipelines would converge on a pair of storage tanks located on the highest point around, Ingombe Ilede. A compact pump house with two diesel pumps sat atop the ridge, which was why I came to Lusitu in 1961. The excavations for the foundations had disturbed nine richly decorated human burials, the remains of ivory traders who had settled in this remote place far from the Indian Ocean during the fifteenth century C.E. My late colleague James Chaplin had cleared the graves as quickly as possible, chivied along by the contractor, who was anxious to finish work. Months later, I arrived to date the site (I got it wrong) and locate any further burials (I found only a couple, both undecorated). It was then that I first witnessed the complex relationships between humans and water, deployed in full array within a few miles.

Few places offered such hydrological contrasts. Some thirty miles (forty-three kilometers) upstream stood the menacing semicircle of the Kariba Dam, a concrete arch 420 feet (128 meters) high. The dam had tamed one of the largest rivers in the world at huge cost to provide electricity for cities and copper mines hundreds of miles away. This vast, expensive structure did nothing for the Gwembe Tonga. Traditionally, village women and girls had collected water in clay pots from the Zambezi and its side streams. They did the same at Lusitu, but the distances were longer, the banks less accessible. Now the Ingombe Ilede pump house would transform the villagers' lives, or so the local government officer told me. Standpipes in each village would bring Zambezi water within easy reach of every household, even if they had to carry it to their homes. Here, timeless practice and modern-day rural plumbing powered by diesel pumps stood side by side. The Tonga now had a reliable water supply, comparable to that experienced by urban Romans and medieval Londoners—as long as the pumps kept working. Even then, young and fresh out of university, I was struck by the palimpsest of relationships between water and people within a tiny geographical area. To the dam builders, water was a commodity, a means to an end. To the colonial government, standpipes were a gift from a munificent authority, a convenient solution to an inconvenient thirst problem, nothing more. But to the Tonga, it was life, one of the ways of defining their society.

Gravity controls river floods, creates runoff, replenishes groundwater with rainwater, and fills lakes and watering holes. In ancient times, water came from three sources—from nearby lakes, rivers, and springs; from groundwater close to the surface reached by wells or through natural water seeps; and from a distance, often with human assistance. Everyone, whether African farmer, Maya lord, Andean noble, or Egyptian official, had to acquire water by obeying the fundamental rule of gravity: Water flows downhill.

The changing human relationship with water throughout history, which I think of as being in the three broad stages outlined in the preface, inspires endless important questions. When did our forebears first harness flowing streams and divert them onto their fields? What special knowledge did they need to do so, and how did they acquire it? Astonishing although it may seem, and as we shall see, such simple diversion works still sustain many farming societies to this day. How did such peoples survive for centuries in extremely arid environments where virtually all water came not from the heavens but from rivers that flowed far over the horizon? What was the role of prolonged droughts and other climate changes in defining sustainability? What happened when more complex societies harnessed gravity to irrigate not just a few fields but entire landscapes? When this happened, who controlled water supplies and crop yields? Where did the power lie—was it vested in the village or in a palace or temple? This is a history as much of common folk as of powerful lords and divine kings, a story of people quietly going about their daily routine in worlds governed by the passage of seasons, by floods, and by complex relationships with the supernatural.

To THE GWEMBE TONGA of the 1960s, water was a precious substance. (It still is.) Every October, they watched massive black thunderclouds piling up over the mountains on the far side of the valley. Flashes of lightning and occasional thunderclaps rent the heavens, but all too often they merely raised the temperature and brought no rain. When the showers came, they could drench one village and leave another a mile away completely dry. Drought was a way of life in the Zambezi Valley, where farming was a high-risk endeavor. I watched the farmers plant with the first showers. A few weeks later, we would walk through freshly planted fields where maize, millet, and sorghum withered in the hot sun. The Tonga were stoic and could at least rely on fellow kin in nearby villages to help them out. I witnessed in unforgettable ways how rainfall and the river defined their lives and those of their crops, and I remembered Ecclesiastes: "All the rivers run into the sea; yet the sea is not full; unto the place from whence the rivers come, thither they return again."[2]

Like the Tonga, we all live with a quiet reality. Water is the principal constituent of all living organisms, making up 65 percent of human bodies, so that without access to dependable water supplies such as a spring or a cistern, any form of activity is difficult; survival may be at stake.[3] When we perspire, we lose water, which has to be replaced as soon as possible. The amount varies depending on where you live. People dwelling in desert climates may sweat off as much as 2.6 gallons (10 liters) a day. But a vast majority of human water consumption goes toward agriculture. And of course, plants are also vulnerable to water deficits, especially in arid and semiarid lands, where up to more than 99 percent of the water absorbed by their roots is lost in the dry atmosphere around them. Cereal crops are especially sensitive to a lack of water to replace that lost through the process of transpiration, equivalent to sweating in animals. Faced with a lack of water to draw from the soil, a plant first suffers from dehydration, then wilts and dies. In many dry areas, rain falls irregularly, usually during a relatively short wet season. Meanwhile, the atmosphere remains thirsty and draws moisture from the vegetation. Water evaporates from the soil and also percolates downward. Even a few weeks without moisture can cause a crop to wither or fail altogether. Water is capricious and powerful, far more masterful than the humans and animals that depend upon it. This is why gravity and irrigation play such important roles in human history.

"Irrigation": The very word conjures up images of lush golf courses and green river valleys teeming with growing crops, of agriculture on a grand scale. We have visions of great irrigation works of the past. We think of finely dressed officials supervising the digging and maintenance of canals. Small armies of villagers gather in the harvest under their watchful eyes. Diligent scribes count every sheaf, tablets close at hand. We've admired, too, many ancient Egyptian tomb paintings of the harvest on noblemen's estates, where scribes assess taxes as the owner leans on his staff. In fact, and contrary to popular belief, almost all ancient irrigation flourished off the historical radar screen in the hands of local communities.

For a while, irrigation defined ancient civilizations in academic circles. A half century ago, a distinguished historian, Karl Wittfogel, wrote what soon became a classic work on water and early civilization. *Oriental Despotism: A Comparative Study of Total Power* described the early civilizations of Mesopotamia, the Nile, and the Indus River of South Asia as "hydraulic civilizations."[4] The substance of his argument was that extensive irrigation works produced huge agricultural surpluses. He believed that water management on this scale required an elaborate bureaucratic infrastructure and economic centralization, which led to the development of urban societies and the first civilizations. Despotic leaders with total power ruled new forms of hierarchical societies based on cities. A form of developmental cycle came into being, where more elaborate water systems and larger food surpluses helped create increasingly hierarchical societies. As food and water surpluses continued to mount, the societies became ever larger and more elaborate. Thus, irrigation was a prime cause of civilization. Wittfogel was a pioneer who wrote at a time when virtually nothing was known of ancient irrigation systems around the world, even in his core areas, China and Mesopotamia. With the hindsight of more than fifty years, we know that his notion of "hydraulic civilizations" masks a far more complex reality. For instance, archaeological surveys have shown conclusively that both the early Sumerians and the Egyptians did not develop elaborate irrigation works for many centuries after cities appeared along the Tigris, the Euphrates, and the Nile. In reality, the state was usually an outsider, its primary interests lying in crop yields and taxation in food and labor. The large food surpluses that filled urban granaries came from dozens of small rural communities producing far more food than the small surpluses they would normally grow for their own uses. I believe that for most of history, the power over water management and food supplies resided, ultimately, in the hands of local communities.

BEFORE THE KARIBA resettlement, the Gwembe Tonga combined subsistence farming based on irregular rainfall with the cultivation of small plots by the river nourished by receding floodwaters in late spring and early summer. Otherwise, they drew water from the river or even drank it out of their hands, just as hunter-gatherers have done since the beginning of time. Like farmers, such people have extremely complex and intimate relationships with water and the way it is distributed across the landscape.

Take the case of the Australian Aborigines. The direct ancestors of the Australian Aborigines settled in their arid homeland at least forty-five thousand years ago. The unpredictability of water supplies makes for an uneasy relationship with the natural world. Aboriginal groups see themselves not as the managers of their environment, but rather as people in a state of constant apprehension and negotiation with it. It follows that knowledge of water sources and their uses is an important part of how Aborigines feel about who they are, and of their perceptions and experiences of their landscape. Over thousands of years, they developed sophisticated adaptations to some of the harshest environments on earth, defined in large part by water sources.[5] Sparse desert populations lived in a state of continual flux. When heavy rains brought standing water and replenished natural wells, the people would move out over wider areas. As supplies evaporated, they would forage their way inward, moving with the diminishing water, until months of drought forced them to camp near the few more-or-less permanent water sources at their disposal. These they used last of all, in an endless, irregular rhythm of movement dictated by rainfall. Fission often followed, then aggregation with small groups of related families anchored to more permanent water sources during extended droughts. Everyone was at the mercy of a capricious and totally unpredictable water pump that could inundate the desert in hours, then dry up, leaving a baked, utterly desiccated world in its train.

To the Australian Aborigines, water is sacred, even elemental, and the source of life. Water is elusive in their generally arid homeland, its sources inconspicuous. It is far more than merely a physical resource. The distribution of water across the landscape defines most hunting and foraging. Aboriginal relationships with water sources of all kinds are both social—delineated by kin and other ties—and spiritual, part of how they know and understand the world. As they travel across their home territories, each band invokes local traditions of ancestors and spiritual beings. These traditions provide knowledge and power, good luck and ways of avoiding misfortune. Rights to water sources and to hunting territories are held in common with other members of kin-based groups. Their territory consists of places that evoke profound sacred

Figure 1.1 *Australian Aboriginal children in Arnhem Land, northern Australia, drinking water from a stream. Aborigines customarily drink with their mouths rather than scooping up water with their hands. (Penny Tweedie/Panos Pictures)*

associations, important memories, and social obligations, a hard-learned landscape of memory and social meaning.

Since the beginning, walking tracks have crossed the arid center of Australia, each following chains of water sources known to generations of hunters. These are Dreaming tracks, or songlines, whose paths are recorded in traditional dances, songs, and tales, as well as in paintings.

A knowledgeable Aborigine can navigate the landscape by repeating the words of songlines. Some lines cover only a few miles; many others, long, memorized paths through the desert that link dozens of water sources—some good, others at best marginal but enough to connect a band with better supplies further along. Water knowledge, of the locations of sources reliable and ephemeral, the telltale signs of such places, and the names of them, passes meticulously from generation to generation. Think of long-established medieval pilgrimage routes, memorizing subway stops in New York, or a bicycle route through San Francisco, and you get part of the idea. Each generation learns what each source can yield, from a deep pool containing thousands of gallons of freshwater shaded in a deep gorge to the few cupfuls that survive in the hollow branches of a eucalyptus tree. In western Arnhem Land, for example, hunters define their territories in terms of locations: "*Indjinaidj*, Point David: waterhole. Sacred site of *jurwa*, the green-backed turtle, metamorphosed in stone . . . Wunuri or Gwunjuri: camp and hunting center . . . Water runs from the 'nine mile' to Warilidj, which is good for hunting possum and collecting wild honey."[6]

Sequences like this—maps, as it were—passed from generation to generation, can juxtapose a sacred, mythological monster, a camp site, and a hunting place. They are linked along a pathway of movement, experience, and memory that changes constantly. The tracks and water sources lie in the midst of a landscape thwart with social and spiritual dangers. A band member is only truly safe if he or she is living with close relatives in a territory to which they belong. It is among such people that the knowledge of how to avoid danger, of how to navigate the landscape safely, resides. Many groups describe their territories by prominent landmarks, including rivers, creeks, and water holes. They also construct their personal and group identities on the basis of ancestral origins, which are mainly expressed in terms of precious water sources.

Knowledge of these places resides with the elders, men and women of unquestioned authority, who have spent much of their lives acquiring firsthand knowledge of the landscape. Their knowledge comes from ceremonies and rituals, from chants, songs, and recitations. Genealogies and local histories are part of the narrative, as are their communications with the ancestral beings and ancestors, who are said to frequent water sources. The elders invoke spiritual protection for their juniors and have authority over access to and use of locations where water is to be found. They decide who receives the knowledge and who joins their company. They select mature individuals, who have passed through the appropriate initiation rites and are deemed capable of maintaining the tight veil of secrecy that surrounds the knowledge.

The elders' knowledge goes back to what is called the Dreaming, or Dreamtime. Awareness of the Dreaming pervades Aboriginal life—the period of creation, a time in the distant past, long ago. Australia was featureless and uninhabited before its transformation by ancestral beings—part animal, part human. The mythic beings of the Dreaming are heroic figures endowed with supernatural powers. Every meaningful activity, every event, leaves behind a residue in the earth, just as plants leave seeds behind them. Streams, water holes, and other places echo the events that created them. Everything in the environment is a symbolic footprint left by the mythic beings, which created the world. They hunted and foraged just like humans, but in so doing, they created salient features of the landscape. In western Australia, the Wagyl, a sacred serpent, meandered over the dry landscape, creating lakes and waterways, among them the Swan River. Other snakes created winding creeks and water holes. A lizard-man's stone ax would carve a gap between two hills. The eggs laid by an emu ancestress would turn into oval boulders.

Stories of water creation abound, always with mythic beings involved. One such story tells us that back "in the first time," when everything was new, a hunting band camped on a mountain in a drought. The band's water sources ran dry. Two greedy members of the group stole the last of the water and ran away with it when everyone was asleep. The elders found the tracks of the two thieves, and the band hunted them down. Spears were thrown; one punctured the water carrier, which began to leak. Eventually, the two men were captured. Strong punitive magic turned one into the first emu, the other into the first blue-tongued lizard. "But a wonderful thing had happened. Wherever the water had leaked onto the plains, there were now beautiful billabongs or waterholes. There was grass and flowers and lovely water lilies and then there were shrubs and trees. And soon the birds came and everyone was happy because there was enough water for everyone." In another story, two Kurreahs, giant crocodiles, seized a man's two wives as they bathed in a spring. He hunted the beasts down and speared them as their tails lashed a hollow in the mud. Then he cut them open and rescued his wives. Each year, the hollow, now the Narran Lake, in New South Wales, fills with water.[7]

Every Aboriginal band thinks of the landmarks and water sources in its territory in terms of what the beings of the Dreaming wrought. The power of the original creative beings surrounds humans on every side. Myths, songs, and ritual are powerful and vibrant forces in Aboriginal life, just as they are in other hunter-gatherer societies. The law of the Dreaming revolves around reciprocity and cooperation between individuals and groups, essential in arid environments.

The rights to live in and exploit a hunting territory depend on at least one member of the band having a connection with ancestors, preferably at the generation of grandparents. The social networks of each band extend over far wider areas, to the extent that people who marry into another band still retain rights to their original homeland through descent and alliance. But to exercise their rights means having some knowledge of both the territory and, most important, its spiritual landscape. These relationships based on knowledge, especially between elders, extend over long distances, much further than the regular face-to-face contacts within a band or neighboring groups. Inherited friendships and other such social mechanisms are powerful factors in the Aboriginal relationship to water, for everyone has obligations to water from the past that extend deep into the future, to their descendants.

The Australian Aborigines have an intensely local identification with water sources, especially springs, in ways that contrast dramatically with the much-larger-scale identities of other societies throughout the world, like Bali rice farmers (or, long ago, ancient Egyptians), whose realms can encompass a water source far away in the mountains, or an entire great river. For the Aborigines, water and water places are constructed socially for the band through the experiences of the elders, who dwell in these places, which to them are also "spiritscapes." The complex and profoundly personal relationships between the Australian Aborigines and their water have deep roots in the past and surely extend back deep into the Ice Age, although as yet we cannot prove this. The power of the Dreaming resides in the waters.

ALMOST ALL HUNTER-GATHERERS live in mobile bands—they can move across the landscape in search of water. Farming communities are literally rooted in place. For them, water is a shared resource subject to the tyranny of sloping terrain. To manage any downslope flow requires careful planning and the excavation of communally owned channels. The routine of irrigation depends on reliable and equitable delivery of water to each field, whether up- or down-canal. Digging and maintaining channels, scheduling watering days and the amount of water for each owner, carrying out emergency repairs in times of flood—these are the kinds of details that require not administrative fiat but complex negotiations between members of kin groups, between landholders up- and downstream, often between neighboring villages. Every irrigation system, however large and complex, works best when decision making is flexible, as much as possible left in the hands of the irrigators themselves. If individual communities are left alone to make their own decisions, much greater stability ensues, with conflicts over delivery—and these are inevitable—settled locally. I once witnessed a violent argument between two Zambezi Valley families over access to a well. By the time the argument was over, a crowd had gathered. Relatives and neighbors joined in with gusto. In the end, the village headman settled the dispute after listening to both sides. In the past, as today, these groups were basically egalitarian societies, where power lay with the ancestors, with kin leaders, and in the powerful hands of public opinion. Irrigation canals might flow for only a few miles, but their digging and maintenance lay in the hands of everyone—the only way the community could maintain sustainable life. Among the Sonjo farmers of Tanzania, for example, a man who refused to participate in canal maintenance paid a fine in goats. It's no coincidence that many such societies survive to this day in various parts of the world. They may be riven with factionalism, sometimes resort to violence, or break apart and re-form in a different way, but in the long term, they enjoy a level of sustainability that is unique in its flexibility.

Many studies of ancient water management concern themselves almost entirely with the mechanics of dams and irrigation canals, with such devices as pumps and waterwheels. In the final analysis, however, as Spanish scholar Miquel Barceló has pointed out, everything depended on the human beings behind the technology.[8] It was people who thought through the general design of an irrigation system, however localized. They looked at the topography, the direction and gradient of the water flow, knowing that everything depended on gravity. They selected a catchment point, carefully laid out the direction and gradient of the supply canals before digging them, sited storage reservoirs if needed, and, if they were in use, located the positions of water mills for grinding grain. The course of the main supply canal dictated the layout of the entire system, for the realities of gravity flow meant that the potential for expanding the system declined steadily the further the water traveled from the catchment point. For this reason, most irrigation systems were local and of small size. For all their human dimensions, they were ultimately artifacts, but artifacts that worked well because of the flexible social institutions associated with them.

Subsistence farmers place great emphasis on kin ties and the inheritance of farming land, and on their close relationships with the guardians of that land, the ancestors. Since most village-based irrigation and water management offers few opportunities for expansion, about the only solution is for people to move and found another settlement some distance away, using other water sources while maintaining social ties. For example, in the Maghreb region of extreme northwest Africa, farmers dwelled, and sometimes still dwell, in fortified villages located just above the main irrigation canal. The irrigation terraces lay below the village in serried rows, the fortifications being a reflection of competition for farming land. Such fortified Berber villages go back to at least the third century C.E. and also occurred in Libya, far to the east. In places like the Maghreb, Islamic Spain, Mexico, and the Andes, the long-term sustainability of water management involved an enormous expenditure of human labor, but once built, raised gardens or terraces could result in huge savings in labor—as their builders well knew.

KARL WITTFOGEL WAS absolutely correct in drawing attention to the importance of irrigation agriculture to preindustrial civilizations, but in fact, water delivery and management had a much broader spectrum. Every early urban civilization acquired water from afar—in the case of Egypt, the Indus Valley, and Mesopotamia, from great rivers whose sources were hundreds of miles upstream. Diversion ditches and earthworks, cisterns, dams, long tunnels that tapped groundwater, and wells—the range of ingenious devices was enormous. No early civilization (except perhaps the Khmer of Southeast Asia, and even they had drought problems) had the benefit of inexhaustible rivers with reliable flow at all times of the year. Some had almost no access to large rivers at all. Irrigation agriculture formed part of the apparatus of sustainability and took many forms.

There were irrigation systems designed to supply water to cities, towns, and villages, whose purposes were mainly domestic, and which in the case of the Islamic world provided water for ritual ablutions. The Romans built aqueducts at enormous expense to provide water for baths, palaces, public fountains, or military encampments. Most irrigation provided water for agriculture, for food production of all kinds. Almost always locally designed and constructed, such systems relied on complex yet flexible water technology and a detailed understanding of the local environment. For example, generations of scholars have assumed that the Romans, with their imposing aqueducts and dams, developed the water-management systems of their colonies in North Africa. In fact, Berber farmers had developed their own sustainable irrigation works several thousand years before the Romans attempted to commodify water management with only partial success.[9] Simple but highly effective Berber irrigation techniques spread into al-Andalus with the Islamic conquest of Spain in the eighth century, and they provided the basis not only for village water management deep in the countryside, but also for the river-based systems that supplied towns and cities. The size of the irrigation system made no difference, for all of them were founded on the basic principle that water flow follows gradients.

As time went on, the scale of irrigation works increased to encompass entire river valleys and landscapes. This expansion was a product of ever-larger political entities and tax-hungry rulers, but also of wars and other miseries that created mushrooming armies of convicts, prisoners of war, and slaves to dig canals, drain marshes, and build aqueducts. Some of the largest irrigation works were those of the Sassanians, in Mesopotamia, during the sixth century C.E., which expanded existing acreage under cultivation at least fivefold. Their diversions of the Tigris were extremely high-risk and involved massive investment in manpower, and their irrigation works soon backfired in the face of political instability, widespread famine, and reduced yields caused in part by rising salinity in poorly drained floodplain soils. More than a millennium was to pass before another nation engineered a watershed and reclaimed land on such a scale, during the Industrial Revolution, with German works in the Rhine Basin.

OUR ANCIENT FOREBEARS revered water. Many small farming societies imagine that their world began in fathomless, still waters. The ancient Egyptians believed that their world had three basic elements: earth, sun, and water. Their existence began with Nun, the primeval waters of nothingness. Then the god Atum, "the completed one," the Creator, preeminent over the cosmos, emerged from the watery chaos. He caused "the first moment," raising a mound of solid earth above the waters. Then the life-giving force of the sun, Re, rose over the land to cause the rest of creation. The pharaohs believed that the life-giving waters of the Nile came from a subterranean stream that flowed in the underworld.

Invariably, water was a gift of the gods. The book of Revelation, in the New Testament, speaks of the "angel of the river of the water of life, as clear as crystal, flowing from the throne of God."[10] In the Old Testament, Psalm 63 links rivers with the Lord:

> *You care for the land and water it;*
> *You enrich it abundantly.*

The streams of God are filled with water
To provide the people with grain,
For so you have ordained it.
You drench its furrows
And level its ridges.
You soften it with showers
And bless its crops.[11]

Water enjoyed sacred properties in many ancient societies, as the source of life. The Holy Quran teaches again and again that "with water we made all living things." Throughout Islamic domains, water devices of all kinds were far more than mere artifacts. A bubbling fountain celebrated the sacred mysteries of water and provided spiritual solace. Allah's throne sat on the waters in the highest world of all. In Central America, Maya lords performed elaborate public water rituals atop great pyramids that also served as rainfall catchments. Their artists sometimes depicted the sacred World Tree, which linked the layers of their cosmos, as a maize plant rising from bands of water. Maize was the sustainer of Maya life. The Waterlily Monster served as the symbol of a watery world of raised fields and swamps. In Bali, farmers still irrigate their terraced rice fields on a schedule dictated by gods and goddesses residing in mountains and lakes high above the fields through a hierarchy of water temples.

Much of this reverence revolved and revolves around the close connections between fertility, human life, and abundant water, around water as life. But where did it come from? Almost all ancient societies believed that water flowed from the depths of the earth. The Nile appeared to emerge from an utterly arid desert; the ancient Greeks considered springs to be sacred gifts emanating from the underworld. Few Europeans believed that rain and snow could provide water, at least until Leonardo da Vinci suggested it. Few people believed him. Eventually, during the seventeenth century, the English astronomer Edmond Halley and, quite independently, the French architect and scientist Claude Perrault used measurements of the catchments of the Thames and Seine rivers to prove da Vinci correct.

Today, we know that water, like living organisms, has a life cycle, a circulation. It falls to earth as rain or snow, some on land, much of it at sea. Much of that descending on land drains into the ocean above- or belowground. The remainder evaporates back into the atmosphere either directly or through vegetation absorbing water from the ground and releasing it into the air as vapor. Vapor from both land and sea condenses into new rain-giving clouds. The world's oceans are, as it were, a huge reservoir, so most of this water is salt; only the water temporarily withdrawn from the ocean is fresh. Only about 8.6 billion cubic miles (36 billion cubic kilometers) of freshwater is on earth. A mere 30 percent of that circulates relatively fast. The remainder is trapped in polar ice sheets or is under the surface, in aquifers or as groundwater. Some stays belowground for thousands of years. The amount of freshwater on earth varies considerably, as does the speed at which it circulates. During warmer periods, evaporation rates increase, there is less ice, and more water circulates. The opposite is true during colder periods.

Freshwater circulates through the world, and through human lives, in a never-ending cycle that operates in a multitude of ways. I've hiked glaciers on New Zealand's South Island where you walk beside fast-moving torrents of cold meltwater running in milky riffles heavy with fine sediment. In Durango, Colorado, you can kayak through the city down a mountain stream, where rapids menace and you capsize with disturbing frequency. Rivers pass through raging cataracts, then flow lazily through wide reaches, the current barely discernible. Rowing in a Victorian skiff down the Thames River from Oxford to London is one of the great civilized pleasures of modern life, for the gentle river does the work for you. Sometimes the contrasts are dramatic. During the dry season in central Africa, I've swum in a deep, calm pond a few feet from the edge of the Victoria Falls, where the long-placid, wide Zambezi plunges abruptly into a narrow gorge over a mile-wide front. The spray, "the Smoke That Thunders," rises hundreds of feet into the air and creates a dense rainforest. Lakes, small ponds, limpid springs, a multitude of streams, and dry wadis, where flash floods can sweep you away in seconds—water passes through our lives in many ways, freezing, vaporizing, shaping the land, creating every kind of ecosystem and environment imaginable.

We have tried to subdue water, to master it, and have come a long way from a simple dependence on lakes, rivers, and springs. Now we

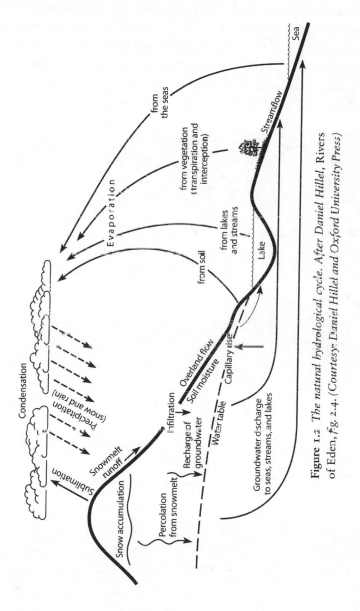

Figure 1.2 *The natural hydrological cycle. After Daniel Hillel, Rivers of Eden, fig. 2.4. (Courtesy: Daniel Hillel and Oxford University Press)*

Condensation

Precipitation (snow and rain)

Sublimation

Snowmelt runoff

Snow accumulation

Percolation from snowmelt

Infiltration

Recharge of groundwater

Water table

Overland flow

Soil moisture

Capillary rise

Groundwater discharge to seas, streams, and lakes

from the seas

from vegetation (transpiration and interception)

from lakes and streams

from soil

Evaporation

Lake

Streamflow

Sea

have probed the depths of the earth, where sixty times more water lies than on the surface. But with deep pumping, there is still not enough water to satisfy our accelerating demands. We are finally at the point where we have to admit that our only weapon is to use our ingenuity to devise ways of consuming less of the most precious finite commodity of all. We're at the point where our existing way of life cannot be sustained by promiscuous pumping of this resource. Profound changes in our attitudes toward, and uses of, water are unavoidable.

CHAPTER 2

Farmers and Furrows

THE EUPHRATES VALLEY: Twelve thousand years ago, early summer. The young girl and her mother sit by the riverbank in the warm afternoon sun, full water skins at their side. Long shadows fall. Spring floodwaters swell the normally quiet river, yellow-brown, swirling in large pools, overflowing onto the sandy plain. A few boulders form natural pools where clear water can be found, newly refreshed by the inundation. The girl kneels at the edge of one calm pond, gathers a drink in her hands, lets the water flow down her arms. She laughs delightedly, casts a pebble into the calm pond. Ripples move quietly to the pond's edge. Gentle waves briefly overflow on a sandbar at the far side. Soon the girl is on her knees, digging with her hands, coaxing the deeper water into a shallow channel. A gentle stream passes down the defile, but soon evaporates into the sand. The girl digs deeper, forming a steeper gradient, and the water flows again.

The mother remembers her grandmother telling her of better days, when oak forests covered the valley floor and rich harvests of pistachio nuts and acorns sustained her family through the winter months. She has heard tales, too, of the great gazelle hunts of spring, when villages from miles around cooperated in the mass slaughter of the small desert antelope as they moved into the fertile valley. Her mother and grandmother have long since departed for the realm of the ancestors, the guardians of the people and the land. Both of them lived through year after year of faltering rainfall, through almost entirely dry years when the forest retreated and the Euphrates flood was at best a shadow of its

former self. Every household turned to wild rye for sustenance. When
the shoots withered in the drought, they started to plant wild seed to
obtain more food. Now they all depend on the crops they grow as much
as they do on wild plants and game. And their entire lives depend on
water, on rainfall, as never before.

How did it begin? When did people first do more than merely collect
water in skins or drink from their hands? Almost certainly, the idea of
diverting water to nourish growing crops is as old as farming itself. Agriculture sounds like a dramatic invention, but it was not. All foragers
knew that plants germinated, grew, and reached maturity. They arranged
their lives around the brief harvests, moving from stand to stand of wild
grasses as they ripened, camping near oak groves when acorns matured.
To plant seeds to amplify existing growth was a natural progression for
people experiencing food shortages. In so doing, they were not trying to
invent a new lifeway, but to increase natural food supplies. Our woman's
aged mother may have lived through the entire changeover, which seems
to have occurred during a mere half century or so. Within a few generations, hunter-gatherers became full-time farmers, and humanity's relationship with water changed radically.

THESE TRANSITIONS UNFOLDED during a period of major climatic
changes. For thousands of years, the Near East had been dry and cold,
at a time when great ice sheets mantled the Alps and Scandinavia.[1] The
Sahara Desert was hyperarid, its margins extending deep into western
Africa. As global warming accelerated beginning fifteen thousand years
ago, ice sheets retreated and sea levels rose. Warmer and wetter conditions descended on the Euphrates Valley and the Sahara. Hunter-gatherers
lived in a proverbial land of milk and honey over a broad swath of country between the eastern Mediterranean coast and southeast Turkey.
Small bands of lightly equipped hunters preyed on antelope on semiarid
grasslands that now extended deep into the Sahara. Plant foods of all
kinds abounded in the oak and pistachio forests close to the Euphrates
Valley. Water was rarely a problem; rainfall levels were considerably
higher than today.

The irregular, but marked, warming at the end of the Ice Age came to an abrupt close about 12,900 years ago, when a huge influx of freshwater from a glacial lake in central Canada cascaded over the northern North Atlantic and virtually shut down the Gulf Stream. Near-glacial conditions returned to Europe for a thousand years, until the Gulf Stream reasserted itself. Now intense drought gripped the once well-watered river valleys of the Near East. As the drought intensified, the hunting populations of the region turned to wild grasses and less palatable foods that were still within easy reach of their villages. Dry years persisted. Soon these alternative foods proved inadequate for a growing population, so the villagers took the next logical step. They started cultivating wild grasses to supplement the natural stands that could be harvested each year.

For better or worse, it was the consequences of deliberate planting that changed human life and our relationships with water. With startling rapidity, in perhaps just a few generations, genetic changes in wild emmer wheat, barley, and rye turned foragers into full-time farmers. Within mere centuries, the fundamental dynamics of human life changed profoundly, largely because people were now anchored to their land in entirely different ways. No longer did they wander over large hunting territories or freely exploit the game and wild plants of such exceptionally rich and diverse environments as those in the Euphrates and Jordan valleys. Their new worlds were smaller, more tightly encompassed by the plots of carefully selected, fertile soils where they planted and harvested cereal crops, and by the patchwork of grazing and browsing areas where they tended their flocks of goats and sheep, and later cattle. Above all, they had to live in places with sufficient rainfall to water their crops or close to rivers and lakes where they could use gravity to divert water to their fields.

If you visited a village of the day, you would find yourself in a maze of narrow alleyways and adobe walls, with just enough space for goats, sheep, or a laden farmer to pass. The only traffic in late afternoon would be women and girls carrying goatskins full of water for their families. The hamlet might look deserted, but you would know that there were people all around you behind the buff walls. A man and a woman arguing passionately, women laughing, the sounds of children playing, and, above all, the steady scrape, scrape of stone grinders preparing the evening meal. Noises would assault you on every side, mingled with the scent of wood smoke, sweating bodies, and rotting garbage. Such villages were small, enclosed worlds. The closely packed adobe dwellings had flat roofs and were separated from one another by small courtyards and the narrow alleyways. Families and kin lived at close quarters, without the flexibility of hunter-gatherer societies, where it was relatively easy to move away when tempers flared and violence threatened. In the close-knit universe of farming villages, you were always tied to kinfolk in your own settlement and further afield, to the surrounding land and the crops that grew there. A tapestry of ancestry and kin ties controlled access to the fields and the manner in which they passed from one generation to the next. Inevitably, there were quarrels over inheritance and landownership in landscapes where rainfall was unpredictable and careful soil selection was the secret to good harvests.

With agriculture, the equation of survival changed fundamentally. Farmers might live near a lake, river, or spring, which would provide drinking water as it always had. They might still eat wild plant foods or hunt gazelle, but their primary living came from cereal crops. A wet year might produce abundant harvests or flood young shoots out of the ground. Droughts and cloudless skies withered crops in the fields. The new equation was brutally direct. People were now dependent on the productivity of their fields rather than on broad hunting territories—and those fields would only produce food if they received adequate rainfall.

The sedentary farming villages lived from harvest to harvest. Their survival depended on the precious food surpluses that guaranteed not only enough to eat but also enough seed for next year's planting. Even the best-watered areas experienced hunger and malnutrition at some point, especially in the bleak months of late winter and early spring, when food was always short. At first, during the earliest millennia of farming, every village had a safety net to fall back on in its herds of goats and sheep and in game and wild plant foods. However, as livestock stripped the land and village populations grew, the expanding fields began to "eat up the land," permanently transforming the natural environment. In the days before farming, the number of people the landscape could support was small. Now, with growing populations after the same wild foods, both game and plants rapidly became scarcer. Gazelle migrations tapered off, and stands of wild grasses disappeared in the face of the digging stick cleaning fields. The potential for hunger in a much drier world was now much greater.

Every hunter knew that water flowed downstream, down steep defiles, through fast-flowing rapids, then idled serenely through meandering valleys. Anyone who could fashion a deep pool by damming a small stream with a stone weir or clay could catch fish. And every child living near a river had spent happy hours digging channels to divert water and create new streams. This collective knowledge of generations must have been the catalyst for the first modest efforts to divert water from rivers and streams into shallow furrows that carried it to strategically placed fields, where the water could be spread with one's feet and hands, with judicious use of digging sticks. Like farming, these first small-scale water diversions had momentous long-term significance. Communities that lived on fertile soils close to perennial, flowing water could free themselves, at least partially, from the tyranny of capricious rainfall.

Thus was born furrow irrigation, the simplest form of water management, a technology so effective that subsistence farmers in many places still use it today. Now humans not only used water but also began to manage it, at first, presumably, along permanent watercourses, but soon in higher-risk situations where streams flowed for only a few days or weeks, or only after sudden storms

THE EPOCHAL DROUGHT caused by the implosion of the Gulf Stream lasted about a thousand years. Around 9500 B.C.E., the Gulf Stream was back to normal. Warmer conditions returned to Europe; much hotter and somewhat wetter conditions spread over the Near East. Mixed farming (agriculture and stock breeding) now flourished in places like the Wadi Faynan, in southern Jordan, about twenty-five miles (forty kilometers) from Petra, famous as a major center of the much later Nabatean civilization.[2] Today, the wadi is a desolate landscape, but there was more water when farmers first settled by a spring on a junction between the surrounding mountains and the main valley. They also founded villages near other springs on the better-watered uplands. Here, they cultivated wheat and barley, using the wadi itself for hunting and herding (figure 2.1).

The settlements on the uplands flourished. Local populations swelled. By the sixth and fifth millennia B.C.E., the main arable tracts had shifted into the wadi itself. A major village rose at Wadi Faynan, at a time when the climate was significantly wetter than it had been in earlier times, or was to be in the future. The village lay by a virtually perennial stream, whose seasonal floodwaters supported most cultivation in a routine that was simplicity itself. Each family planted its crops in the damp soils on either bank of the stream left by the receding floods. The same practice was commonplace among early farming communities along the Nile and persists along rivers like the Zambezi into modern times.

About six thousand years ago, the climate became progressively drier throughout North Africa and the Near East. In the Wadi Faynan, the landscape was now increasingly steppelike. Virtually all farming activity was now concentrated at lower elevations. The shift came at a time of profound social change throughout the Near East, with, among other developments, the first appearance of metallurgy, a massive expansion of trade networks, and the appearance of the first much larger communities, towns if not near-cities, most of them near major rivers. The wadi contains rich copper deposits, which soon attracted attention, especially after 3500 B.C.E. With the increased desiccation, the wadi's farmers could no longer rely on well-watered locations. Instead, they began to develop new strategies for living in a much drier environment. They turned to floodwater farming.

THE ELDERS LIVE with their ears cocked during the rainy season. Day and night, they listen to the wind, watch gathering clouds, and are alert for the soft rattling of water-washed stones in the dry streambed in the nearby gully. Sometimes, the noise is subdued, transitory, as a brief freshet cascades downstream for just a few minutes. Then there are the late afternoons and evenings when massive, rain-filled clouds crowd over higher ground. A flash of lightning, crashes of thunder—only a few heavy drops fall on the village, but the men are already at the crude stone dam and terrace walls they reinforced during the hot summer months. A loud concatenation of rumbling boulders and a tumbling surge of sand-colored storm water wash against the dam and its terraces as the torrent submerges them. The elders watch silently, as their relatives pile boulders to reinforce weaker terraces. Minutes later, the clouds lift upstream. The mellow light of the evening sun sparkles on the now-falling waters. Long into the night, the villagers watch the flood. As the check dam—built to collect floodwater—reappears and the stream recedes, young men prowl immediately downstream, plugging leaks with stones and clay. Behind the crude barrier, a large pool of floodwater stands ready to fill the nearby stone cisterns and water the nearby fields.

STUDYING ANCIENT IRRIGATION systems is never easy, especially in places with a long history, like Wadi Faynan, where the irrigation works of centuries lie atop the small-scale efforts of the earliest inhabitants. All that remains are field boundaries, stone alignments from long-disused terraces, and large boulders that formed the foundations of check dams. Thousands of years of cultivation have swept away small dwellings and goat enclosures, the delicate traces of human occupation, such as charcoal-filled hearths, that enable us to date farm works. Landscape archaeology is a laborious task of field survey and mapping using GPS devices, computers, and, in the case of more elaborate field systems, aerial photographs. This costs money and consumes many field seasons. The Wadi Faynan Landscape Survey involved a multidisciplinary team from more than seventeen universities.[3]

The landscape survey was able to decipher the imprint of farmers on Wadi Faynan during the fourth and third millennia B.C.E. The villagers lived in irregularly built stone dwellings close to field systems marked with stone boundaries. They built check dams and crude terrace walls across the shallow floors of tributary wadis. In this way, they trapped and collected floodwater in circular and oval storage cisterns up to about twenty inches (fifty-one centimeters) deep. In some places, they laid out terraced fields along the direction of water flow. Simple but apparently effective for many centuries, the Wadi Faynan check dams and terraces involved no elaborate organization. Nor did they require social mechanisms for deciding who got water for their fields and how much. But their simple efforts to divert water onto fields were the prototypes of much larger and more tightly controlled irrigation works in later times. Like floodwater farming, check dams persist to this day, especially in Africa, where one can achieve some understanding of the basic yet highly effective ways in which village societies manipulate water and the acreage it feeds. Village irrigation remained a foundation for some of the most complex preindustrial civilizations on earth.

I'll never forget driving northward through northern Zambia in late September and October. The gravel road passed through mile after mile of savanna woodland, dipping occasionally for a shallow, dry watercourse. It was a dull drive at the best of times, especially at the end of the dry season, when the farmers burned off their gardens. Dark, low clouds of ash-filled smoke clung to the treetops, masking the sun, swirling through clearings. A thin cloud of wood ash covered our tents each morning and winnowed into our coffee. Everything was pale and washed-out, even the sky, as the farmers prepared for rain.

Poor soils, uncertain rainfall—one expert has estimated that about 40 percent of sub-Saharan Africa's soils are marginal for subsistence agriculture. The classic model of such African farming is swidden, or slash-and-burn, farming, well documented in a series of classic anthropological studies conducted before and after World War II.[4] Slash-and-burn involves just that. Village farmers clear savanna woodland, burn the resulting timber and undergrowth, then hoe the ash into their new gardens. After a number of years, sometimes as few as two or three but occasionally after as many as seven harvests, crop yields fall rapidly. The farmers move on, leaving the exhausted soil to lie fallow. They clear new gardens, often as a series of concentric rings ever further from their dwellings. After a generation or so, the village moves, to be closer to growing crops. If there is plenty of land, it may shift some distance into

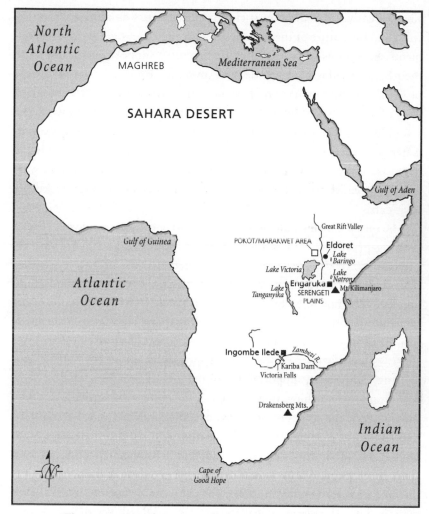

Figure 2.1 *Map showing major locations in chapters 1 and 2.*

deserted bush. In most cases, however, and especially in places where population growth has crowded the landscape, the village will move in a gradual circle, as farmers take up land that has lain fallow for several years.

If you map the population of east Africa, you find that the density varies greatly, depending on such complex factors as altitude, rainfall, soils, and the nature of the societies that occupy the very diverse environments. Enormous tracts of open, dry country support pastoralists and their herds; swidden farmers depend on seasonal rainfall over vast areas of the savanna; bananas are a staple in low-lying areas with abundant rainfall. One would think that the density of human settlement would be a simple consequence of generally adequate rainfall or shallow groundwater. In practice, however, many other factors are involved—people move as a result of overcrowding, because of endemic raiding and warfare, or because tsetse flies have expanded their range and are killing their cattle.

The term "shifting cultivation" covers an enormous range of African farming practices. Generations of anthropologists and historians assumed that African agriculture had changed but little over the centuries, that slash-and-burn farming, with its generally low crop yields, had long been the norm. Nothing could be further from the truth, for many innovations took hold over time, including many elaborations on swidden farming. The most innovative areas, especially in east Africa, are those where altitude, rainfall, and topography allow permanent, not shifting, cultivation, supported by all manner of practices, among them intercropping, mulching, ridging, manuring, and furrow irrigation. These areas include the slopes of northeastern Tanzania's Mount Kilimanjaro and Mount Meru, higher altitudes in that region, and the Kenya highlands. One staple crop is bananas, which have been cultivated for at least a thousand years, and probably longer, but it is cereal agriculture—the growing of sorghum, finger millet, and, in recent centuries, maize—that demonstrates the remarkable effectiveness and flexibility of simple irrigation in environments with diverse topography and uncertain rainfall. Such agriculture involves investments in land and water management that go far beyond the need to survive from one harvest to the next, capital investments that masquerade in the technical literature behind a rather strange expression: "landesque capital."

The term is misleading, because the kind of irrigation and water management described in this chapter does not involve massive, one-time investment by large numbers of people. Rather, long water channels and field systems develop gradually, the cumulative result of household and village labor over sustained periods of time. Agriculture, constructing irrigation works, and maintaining both channels and fields are activities that blur one into the other—land and water are, effectively, part of the same resource. There are none of the armies of laborers associated with later Mesopotamian and Chinese water projects. Village irrigation in east Africa is a gradual, long-term investment in the landscape, which develops over many generations and is controlled by families, households, and kin groups, who contribute labor to enable an irrigated environment to come into being. The return on investment is cumulative and slow.

These were very ancient practices that continue today. Long before the arrival of Europeans and colonial settlement, the irrigation societies described in this chapter were self-sustaining for many generations. The best-known examples are in east Africa, but there are furrow farmers in Ethiopia. They also once flourished in the highlands of eastern Zimbabwe. And huge complexes of terraces, field boundaries, and water channels can be seen in South Africa's northern Drakensberg Mountains.

In the past, as there are today, there were, effectively, islands of more intensive agriculture in east and southern Africa, some of which supported fairly large populations for long periods of time. These societies were less fragile in the face of drought and social unrest than surrounding groups, who relied on herding and dry agriculture alone. (Dry agriculture relies solely on rain.) Such enclaves, with their simple irrigation channels, offered protection against hunger. In 1940, a colonial officer writing in a famine year remarked of the Marakwet of northern Kenya (described in chapter 3) that "they have been saved by their furrows."[5]

What is remarkable is that nearly all of this irrigation farming flourished over long periods of time without being controlled by powerful chiefs or hierarchies of nobles and officials. Instead, village farmers controlled their own destinies, largely because of land rights that were vested in lineages and clans—in the intricate ties of kin that linked family to family, village to village. The same lineages and clans dug, owned, and maintained water channels and controlled water distribution, for land and water went hand in hand and were, effectively, the same thing.

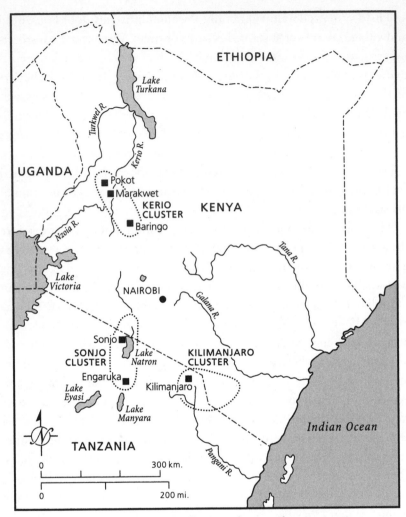

Figure 2.2 *Centers of intensive agriculture and irrigation in east Africa.*

The one could not exist without the presence of the other. It's no coincidence that the same social mechanisms that settled decisions about land and water management also played a central role in marriage negotiations. Land, water, and society were as one. Africa's islands of agriculture existed in isolation, but were closely tied to the worlds of their neighbors through all manner of formal and informal contacts and exchanges. Fortunately, research on more than five centuries of irrigation tells their still-in-progress story.

THE GREAT RIFT VALLEY cuts through east Africa with a giant tectonic slash—deep, with precipitous escarpments in places, and blessed with a few lakes and huge tracts of arid, semiarid, and often very hot landscape. This was one of the playgrounds of the very first humans more than 2.5 million years ago. It is one of the very few places where you can still get an impression of the Africa of ancient times. To witness a wildebeest migration on Tanzania's Serengeti Plain is to experience a nearly vanished Africa, where a magnificent bestiary and legions of predators once caused farmers and herders to be ever-watchful. Altitude means everything here, for rainfall is more abundant, if still variable, on higher ground, whereas the wide valley floor is dry for most of the year. An unlikely place for farming at the best of times, you might think, yet a few people managed to thrive off irrigation agriculture alone here, or to combine it, at least partially, with dry farming. This is the story told by Engaruka, home to the earliest known furrow farmers in sub-Saharan Africa.

The site of Engaruka lies by the eastern escarpment of the rift, in northern Tanzania, at an altitude of 3,280 feet (1,000 meters) above sea level, a low altitude for this part of east Africa.[6] It's a dry and dusty location, with unreliable, variable rainfall, a place where evaporation rates are high, the heat is often extreme, and the rainfall is no more than about 16 inches (406 millimeters) in an average year. Archaeologist John Sutton was lucky here, for the ancient irrigation works lie exposed and intact. Behind Engaruka, the escarpment climbs to around 6,500 feet (2,000 meters) or higher, to highlands with two or three times the amount of rainfall that occurs in the rift. Successive groups of pastoralists have herded cattle, sheep, and goats on the highlands, the most recent of them being the Maasai. No farmers ever cultivated the higher ground, but the runoff that descends the escarpment through deeply cut gorges allowed Engaruka's inhabitants to farm a landscape that would never have supported dry agriculture even in the best of years.

Three spectacular gorges incise the base of the escarpment over 5.6 miles (9 kilometers) behind Engaruka and bring water to the lowlands. Only the Engaruka River brings water year-round, a shallow, fast-moving stream. The other watercourses are more unpredictable. Cumulatively, however, they have created large outwash deposits of tillable soil. Their waters enter the valley at a commanding altitude, with the terrain falling away sharply on either side, in the case of the Engaruka River for more than 1.2 miles (2 kilometers). Permanent water and the topography combined to create an exceptionally favorable location for irrigation farming in a landscape where any form of cultivation would otherwise have been impossible.

About 4,942 acres (2,000 hectares) of ancient irrigation canals and fields lie at the base of the escarpment and spill over onto the plain. The farmers faced serious obstacles in a terrain dominated by stone. Thousands of stones had to be removed before a single crop could be cultivated. Many of them went into the crude stone walls used to demarcate the small fields. Their owners leveled the plots with simple revetments and shallow terraces. They used still more rocks and scree to line their canals and feeder channels. What, then, to do with the remaining stone from the fields? The solution lay in compact, well-built square and angular cairns that stood up to six feet (two meters) high and dotted the fields, each with a rubble core. In this way, as much land as possible could be cultivated without the backbreaking labor of carrying heavy baskets of stones considerable distances. Circular enclosures of thick, rubble-filled stone walls up to thirty-three feet (ten meters) across also lay among the fields. One would have thought that these were protected houses, but they were probably used for cattle, who were stall-fed, there being inadequate pastureland anywhere nearby. The beasts provided not only meat and milk but also manure that could be spread on the surrounding fields. Their owners dwelled in seven large villages situated on the scree of the escarpment, above the irrigated area but close to adequate water supplies for domestic use. Had you visited the fields at night, you would have found men and boys watching the cattle, perhaps huddled by fires, alert for marauding lions.

The combination of cattle herding and irrigation agriculture at Engaruka is by no means unique. There are historic examples of such practices among modern-day people like the Sonjo, who live on better-watered land about sixty-two miles (one hundred kilometers) to the north (see chapter 3). Some farmers in southwestern Ethiopia live in stone-walled villages in a region of unpredictable, seasonal rainfall. They draw water from stone-lined wells and irrigate fields in narrow valley bottoms with narrow furrows linked to perennial streams, which are, however, vulnerable to severe droughts like those of recent years. Like these people today, the Engaruka farmers probably cultivated sorghum, a common cereal crop in sub-Saharan Africa, which has the advantage of being both drought resistant and tolerant of irrigation. They may also have grown finger millet, which has excellent storage qualities. John Sutton believes that this may have been grown without irrigation, then stored both as a famine food if need be and for brewing beer. Various pulses, such as beans, served as secondary crops and provided valuable nourishment for the soil.

Successful farming depended on gravity, on water fed through shallow furrow canals from springs or mountain streams. Arid and stony Engaruka is an extreme farming environment, which is perhaps why the inhabitants lavished such care on the details of their fields, enclosures, and cairns. The first settlers must have farmed on a small scale, with their canals and fields confined to the easiest areas to cultivate, immediately downstream of the Engaruka gorge. As the population increased, the people expanded their fields onto higher, rockier ground.

The farmers invested cumulatively in their waterworks. The main feeder canals ran from the highest possible points in the gorges along the base of the escarpment. Temporary stone walls and other devices diverted water from the streams and must have been replaced at frequent intervals, especially after flood surges. Using stone linings and carefully shaped embankments, the farmers steered the canals around small hills and through their field systems. The gradients of the feeders were sedulously level, both to control the water flow and to prevent erosion and breaching of the sides. You can imagine the diggers shaping the floor of the canals with hoes, digging sticks, and their hands, shaving the ground as the first water flowed down the new defile. Even a few fingers' worth of soil would make all the difference. The stakes were high, for careful attention to the gradient also ensured that water reached the largest area possible. Sutton mapped the canals and found that most of them fell at very shallow gradients. No sophisticated engineering was involved, just a combination of experience and trial and error.

The longest of the Engaruka feeder channels extends over as much as 1.8 miles (3 kilometers) and is up to six feet (two meters) wide, with, of course, a much narrower bottom. A gravel embankment supports the artery as it descends the escarpment. At the foot, it swings abruptly at right angles toward a small, isolated hill. The builders had to maintain the gradient, so they carried the canal across a narrow valley on a stone embankment, effectively an aqueduct up to nine feet (three meters) high. At the hill, the canal divides into contour furrows that follow each side of it, in places apparently once bolstered with wooden scaffolding where hollowed-out logs carried the water even further. Hollow logs were also used elsewhere in efforts to carry the water further still through an irrigation system that was becoming seriously circumscribed.

Engaruka's irrigation system could never nourish more than a relatively finite area, for gravity-fed water could only flow so far along even the highest and longest canals. There were limits to the number of people who could be fed off the tightly packed field systems, even with the careful use of cattle manure and meticulously supervised crop rotation between cereals and pulses. The farmers' options were few, for they could not afford to shorten the fallow period in their fields, which would have rapidly exhausted the soil and caused uncontrollable erosion. At lower elevations, redeposited soil could flow down onto terraced fields, with catastrophic results.

The irrigation works were under even more stress owing to a decline in rainfall at higher elevations relatively soon after first settlement. As the flow from the Engaruka, the main water source, faltered, the farmers tried to adjust their canals. They did this in ingenious ways that give an impression of adept flexibility. More likely, however, they were desperate. As less and less water flowed off the escarpment, the people dug feeder canals off lesser streams and even off some that flowed but sporadically. These works came into use when rainfall was somewhat higher than today. Some now-only-occasional streams still carried water for considerable periods after highland storms, in great enough quantities that field systems could depend on them.

At Engaruka, generations of cumulative investment faltered after initial success. The first settlers had cultivated a limited area with relatively simple irrigation methods. The villages prospered, and the population grew, perhaps at a time when the annual rainfall was slowly declining. The farmers responded to less predictable river flows by expanding their canals and furrows to the maximum extent possible, until about 4,950 acres (2,000 hectares) of intensely cultivated land depended on water from higher elevations. Even a slight decline in water volume made a significant difference, especially when every square foot of cultivable soil was planted to the full. Even with manuring, soil quality declined as the farming season shortened.

Eventually, there came a point when the Engaruka villages were no longer viable and the people moved elsewhere. Judging from the state of the surviving channels and field systems, the move was a relatively abrupt one. Engaruka, with its total dependence on irrigation agriculture, had become too specialized, too closed in, to survive independent of other settlements. The inhabitants moved away, but their fate remains unknown. Conceivably, they moved to the Sonjo region, where several villages still flourish on the escarpment, west of Lake Natron. Here, the inhabitants cultivate fields fed by local springs and also grow sorghum and finger millet by irrigating nearby river valleys. They do not depend exclusively on irrigation, as the people did at Engaruka. The rainfall is higher, so the Sonjo's irrigation works tend to be less well developed than those of their former neighbors. Nor do the farmers own cattle, which were an integral part of the closely organized Engaruka regimen. But the similarity of the compact villages and their agriculture hints at some as-yet-unstudied historical relationship. It was not until the twentieth century that a new community of irrigation farmers came to Engaruka, to fields further downslope than the ancient settlement.

For thousands of years, channels, furrows, check dams, and terraces have helped define the close relationship between people, their land, and their water. As Engaruka reminds us, when the channels and furrows falter, so, too, do the societies that depended on them. The only recourse is to try again elsewhere. As the east African societies described in chapter 3 remind us, it's a fatal mistake to put all your eggs into one proverbial basket or clay pot.

CHAPTER 3

"Whoever Has a Channel Has a Wife"

"Resuming our way we were almost baffled in our attempt to descend, owing to the excessive steepness of the way . . . I here noticed the employment of canals for irrigation . . . many of them being conveyed with surprising judgment along the most unexpected places . . . The men camped beside one of the artificial canals employed to bring water from a great distance, to irrigate the ground at the base. In camping here, we found we had simply delivered ourselves into the hands of the Philistines. The natives at once put the screw upon us to extort a large hongo [bribe]. Seeing us hesitate, they quietly retired, and the water with them—for they could easily divert it in its upper course. This was quite sufficient to produce the desired effect. We humbly paid up; and immediately, as if a modern rod of Moses had struck the rock, the water began to flow."[1]

The rod of Moses: Victorian explorer Joseph Thomson led a Royal Geographic Society expedition through east Africa's Maasailand in 1883–84. On the Elgeyo Escarpment, northwest of Lake Victoria, he learned the power of the furrow, or rather that of those who control it. Thomson spent but one night among the irrigation works, then moved on to Lake Baringo. He also tried and failed to climb Mount Kilimanjaro, in northern Tanzania, but constantly remarked on the spectacularly varied east African terrain.

Thomson traveled through arid grasslands and up steep escarpments and visited higher altitudes where heavy rainfall nourished lush vegetation. Living conditions and ways of life changed dramatically within a few miles or a thousand feet of altitude. Thomson must have passed by other irrigation schemes, though it was only when he camped in the midst of one that he mentioned it. Irrigation has a long history in the arid and semiarid regions of east Africa. We're not talking here about fields planted in damp soils left by receding river floods, as was the sporadic case in places like the Rufiji River, in southern Tanzania, but about larger-scale efforts that involved diverting streams and digging long furrows. Three clusters of such irrigation works flourished in Kenya and Tanzania before the arrival of Europeans (figure 2.1).[2] One encompasses Engaruka and the Sonjo, in northern Tanzania's Great Rift Valley between Lakes Manyara and Natron, already described. A second cluster surrounds Mount Kilimanjaro, where the Chagga people and others draw water from mountain streams and storage ponds with an extensive furrow network to cultivate bananas and cereals such as sorghum. Some of these communities used irrigation to grow surplus food to supply slave caravans during the mid-nineteenth century. The third cluster lies in the spectacular Rift Valley landscape of the Kerio region of northwestern Kenya. Here, furrows draw water from streams at 9,000 feet (2,743 meters) and carry it downslope to more than 6,000 feet (1,829 meters) above sea level. Also, there was once extensive furrow irrigation near Lake Baringo, to the northwest, again to supply slave caravans.

The word "irrigation" conjures up misleading images of wider, checkerboard landscapes bisected by canals large and small, where watchful officials scrutinize every planting, every crop. The reality for furrow farmers like the Marakwet and Pokot farmers of the Rift Valley is very different. Here, dozens of subsistence-farming communities still water their fields using the same kinds of general practices as those used by much earlier farmers in Jordan's Wadi Faynan and at Engaruka. They use their furrows and canals not to grow food surpluses for distant lords, but as a reliable safety net in times of drought and possible food shortages. And there's another daily necessity: Thirsty cattle need water at least once a day. It's worth taking a closer look at Marakwet and Pokot water management, a notable example of the effectiveness of small-scale irrigation.

THE WATER STILL flows steadily down narrow earthen furrows in northwest Kenya, ebbs around hillside boulders, through hollowed tree trunks.

You climb up steep hillside paths, follow the gradually sloping defiles from the river high above through a patchwork of small gardens and growing crops. In places, men with iron hoes are clearing silt and piling it up downslope from the carefully graded channel. A small group of villagers gathers around a recently felled and hollowed-out tree trunk. They are arguing about its length and where to divert the water as they replace the older wooden aqueduct. As their crops ripen, and just as they have for centuries, the farmers gather to argue over water allocations for crops and cattle. They still depend on consensus and public opinion to manage water brought from a distance, just as farmers have done since the earliest days of village farming.

In Kenya's Eldoret region, water cascades down steep hillsides from the Cherangani Hills into the Kerio Valley, more than 3,280 feet (1,000 meters) below. The Marakwet people of the Kerio have developed an impressive expertise in taming these fast-flowing waters.[3] The farmers themselves see nothing unusual in irrigation, in the many miles of canals, furrows, wooden aqueducts, and scaffolding that bring water to their fields. They call it *keir ber,* "to water the land." They also speak of *kekwat ber,* "to level the land," the most important element in any form of gravity-fed irrigation. Irrigation is part of the daily fabric of Marakwet life.

When the Marakwet or their predecessors first began furrow agriculture is unknown, but judging from Engaruka, such practices date back many centuries. Perhaps they developed as a response to growing population densities, but in the case of the Kerio Valley, irrigation may have come into use as a convenient way of growing crops and then resulted in locally much higher population densities.

Marakwet farmers live in a world of highly variable rainfall from one year to the next. Each year, too, climbing temperatures lead to high evaporation rates. Here, farming would be a high-risk proposition without irrigation. From the beginning, irrigation was part of the Marakwet subsistence-farming system—a specialized part, it is true, but valuable because it made farmers less dependent on unpredictable downpours.

Once they had invested heavily in canals and furrows, each family and household became closely tied to the land in ways that were not necessarily the case with farmers who cleared plots, cultivated them each rainy season, then moved on to clear new land when the soils became exhausted. Along permanent rivers, larger communities with stable populations flourished for many generations, as the scale of the canals and furrows expanded. The ties between the people and their irrigated land were reflected in much closer links between different families and descent groups, the kin that determine inheritance of fields. Landholding descent groups constructed the channels, with everyone contributing labor. If you needed to get water to your land, you had to contribute labor to constructing and maintaining the irrigation works. Every member of the community was involved—they had to be. There was no problem recruiting labor for canal digging or maintenance; your genealogical status gave you a share in specific furrows. This guaranteed access to land and water to all members of society.

The social equation was simple enough. Those who fulfilled their obligations of communal labor that helped keep the furrows in good repair were guaranteed water. Those who did not were excluded, despite continual negotiations over exactly where water went. Digging and maintaining the furrows was not a hardship but a festive activity, despite the labor involved. We can imagine a group of men with hoes, digging in a row, singing as they swung their arms, the soil swept away with deft flicks of their tools. The work would begin in early morning, cease during the heat of the day, then resume as the shadows lengthened. No one was in any hurry. They would work perhaps two or three days a week under the direction of several elders, who scraped and contoured the bottom of the new channel with the careful eye of long experience.

The Marakwet were far more than just hill-furrow farmers. They had a long tradition of herding, especially in the lowlands. Cattle were a source of pride and wealth. Acquiring livestock was an opportunity to move onto the plains and become full-time herders. Both anthropologists and colonial officers remarked that the Marakwet preferred pastoralism over the hard work of farming, but this is a misleading stereotype of a people who are very flexible in their relationships and economic activities. In recent years, population growth has made it increasingly difficult for the Marakwet to maintain such flexibility, especially in an era of overcrowding and constant land disputes.

The first furrows must have led off the lower reaches of the rivers and streams, dug by small villages, even individual households. As the population increased and the farmers began to construct much larger and longer furrows higher up the rivers, the labor involved expanded. Someone had to survey the route with meticulous care and anticipate exactly where water would flow around large boulders and down slopes. There were gullies to cross, embankments to be constructed, potential landslides to be factored in, all of this apart from the brutally hard labor of clearing bush and building passages through rocky scree, shared by communities the length of the waterway. There was always the risk of failure, of months of work being abandoned to the bush when the furrow was unsuccessful because of too-loose soil. Anthropologist Wilhelm Östberg of Stockholm's National Museum of Ethnology describes how, in the 1970s, five elders of the Talai clan surveyed the route for a new eight-mile (thirteen-kilometer) furrow from the Embotut River that would not only irrigate the clan's fields but also provide water for four schools, two churches, a dispensary, a trading center, and all the households along the way. When the survey started, the people said that the elders had gone to negotiate some marriages. "When water comes home, it is a marriage. Whoever has a channel has a wife." They compared owning a furrow to possessing livestock—as a way of attracting women. Marriage prospects improve. "You will not have to talk long to a girl before she agrees. Women want a place with water, where mangoes grow."[4] By the time the furrow was well under way, ninety-eight people were laboring on it, up from the original team of twenty-two. People living many miles from the river desperately wanted water and were prepared to dig two or three days a week, despite interruptions caused by attacks from neighboring Pokot bands. The channel building took years.

The larger the scale of furrow agriculture, the more organization is required, which in the case of the Marakwet came from their kin-based landownership. The same pattern of cooperation applied, whether the Marakwet were arranging a marriage, clearing new land, preparing for defense, allocating water, or resolving disputes, and it still does today. Repairing a canal can take several weeks, as people work for a few days, some hours, and the number of men working ebbs and flows. But eventually the work is completed, as it always has been.

EVERYONE HAD SOMETHING to say that hot African afternoon in the meeting that I attended in a Gwembe Tonga village deep in the Middle Zambezi Valley in Zambia—an argument to make, a case to be advocated, a pearl of experience or wisdom to share. We sat in exquisite discomfort on wooden stools or the ground, sweat dripping from our brows. At times, everyone tried to speak at once until the loudest voice prevailed. The meeting concerned the contested ownership of a plot of cultivated land nearby, a dispute between fellow kin that affected several other land-owners in the area. What began as a concatenation of angry voices turned into a litany of reminiscences and stories about the recent ances-tors of those who claimed the land. The only time the gathering became silent was when an elder spoke. Then everyone listened, apparently out of respect for a voice of experience, even if they didn't necessarily con-sider it to be one of reason. Eventually, the heat prevailed. The argument mellowed into a conversation that ebbed and flowed across the crowded meeting. As the tempo of the voices slowed, two elders spoke at length. They were eloquent, long-winded but calm. With the benefit of prece-dent and hindsight, they mediated the dispute in a way that left everyone, if not completely satisfied, at least with something gained.

Attending village meetings is part of an archaeologist's life in sub-Saharan Africa, but I'd forgotten just how important they are in commu-nities where decisions are made collectively, with a careful ear attuned to public opinion. There may be shouting, tempers raised, and eloquent argument, but in the end, the decisions are made and accepted by the group. It may be messy, at times chaotic, but the results are remarkably egalitarian. The crux of many disputes at the meetings I attended was land, for these were farmers who practiced dry agriculture. In many so-cieties like the Marakwet, however, the same process of decision mak-ing extends to people who share canals and their water. The gatherings that centered on water in Engaruka are long lost to historical oblivion, but the ways in which the people there distributed this most precious of resources must have involved endless argument and negotiation.

Almost everywhere, every initiated male in the community has a voice in village affairs. The actual process of decision making varies con-siderably once all voices have been heard. Some groups, like the Sonjo of northern Tanzania, described below, who may number descendants of Engaruka farmers among them, leave the decisions to a council of elders, who claim direct descent from Khambageu, a supreme being and cultural hero whose position is akin to that of God. In earlier centuries, powerful rituals and intercessions with revered ancestors and divine beings were always part of the process, but these have largely passed into history. Ritual, communal ownership of the land, and collective use of water were always intricately entwined.

There is much more to this than councils and collective decision making. Access to water, land, and food in times of need is a fundamental daily reality and a critical factor in arranging marriages. The Marakwet follow marriage rules that allow polygamy and make everyone marry outside their own immediate settlement, even excluding neighboring people. They know that the more numerous they become, the better use they make of land and water and the less vulnerable they are to marauding wildlife and raids. Furrow farming is relatively secure, especially in poor rain years, when there is hunger among those who depend on rainfall for their crops. Outsiders can join the irrigation system, for the extra labor is often welcome and there are well-established rules for allocating both land and water. Over generations of at times irregular growth, the furrows have expanded and the system has constantly reinforced itself. Inevitably, though, with population growth in recent times, it has come under stress. There has been an explosion in lengthy disputes over land and water, as well as what can only be called deceitful dealings. Stealing water at night to water one's field is a common offense, especially in drier years. Nevertheless, the system is flexible enough to accommodate historical irregularities, while upholding the ancient legal and moral code upon which Marakwet farming depends.

For all their apparent security, the farmers rely on livestock, especially goats and sheep, as currency, acquired and dispersed as part of the daily traffic of living—for marriage negotiations, as a way of buying water rights, for fulfilling social obligations, and of course, for their meat and hides. The small stock requires little maintenance and feeds on crop residues during the dry season. Animals were once part of elaborate exchange networks that linked highlands and lowlands and provided a means of trading grain surpluses and other commodities with other groups—an additional form of security against hunger.

Marakwet irrigation developed over many generations, as farmers, hemmed in by neighbors, began expanding away from major rivers on the west side of the Kerio Valley. In the end, more than forty channels watered nearly 9,900 acres (4,000 hectares) of farmland. There were distinctive building blocks for the system—the central role of descent groups made labor available; decision making involved everyone. The flexibility and expansive nature of these kin ties made it possible for the Marakwet to build elaborate and lengthy irrigation channels. As Wilhelm Östberg remarks, "it was not only the Marakwet who made the furrows, but also that the furrows were instrumental in forming the Marakwet."[5]

MOST OF THE ancient rituals that lay behind furrow irrigation have long vanished in the face of Christianity, so we know little about them. Fortunately, the Sonjo, who live in the Lake Natron region and may have had some relationship with the inhabitants of Engaruka, preserve some of their traditional beliefs.[6] These farmers herd goats and sheep and cultivate cereal crops, relying on both seasonal rainfall and water from permanent springs and streams. Carefully maintained channels pass water onto the irrigated fields in a flat alluvial valley. Just like the Marakwet and the Pokot, the Sonjo use irrigation to amplify their farming as an important cushion against famine. The streams periodically flood the valley and provide moisture for the fields, but the springs, each considered sacred, flow year-round. A council of seventeen hereditary elders, which meets almost daily in one of the village plazas, controls irrigation and water allocation.

Sonjo farmers dwell in close proximity to one another, in a harsh environment of constant water shortages where the potential for conflict abounds. For this reason, they value calm, judicious behavior, which is why the council of elders is of such importance for adjudicating disputes over the contentious issue of water rights. The assembled elders decide who gets water, for how long, and when. They also levy fines for failing to work on communal tasks and for quietly stealing water at night. The penalties are invariably modest, with repeat offenses commonplace. This builds a nice flexibility into a system of water allocation that favors both the elders and other more privileged members of society. It is as if water stealing is a form of institutionalized conflict.

Every decision made by the council is in the name of the mythic founder and leader Khambageu, who is said to have established the perennial springs. Khambageu died and now resides in the mountaintops, the all-influential Supreme Being. It is he who opens the "sluices of Belwa," a legendary village in the heavens invoked in every community's water prayers. Each year, ancient ritual still entwines with the routines of water management. Between October and December, the solemn Mase festival unfolds at each village in succession just after the irrigated fields are planted, for in coming months, water will assume great importance in village life. (The name Mase has no known significance.) This is when Khambageu visits each community in turn. A trumpet sounds in a closed hut near his shrine in the plaza and proclaims his arrival. The villagers, dressed in traditional leather garments, gather near the fenced-off temple at dawn and offer goats and other gifts to the priests for the visiting divinity. As the sun climbs overhead, young men assemble in the plaza. They form a circle, singing and dancing, rotating counterclockwise toward Khambageu's shrine. After a break at midday, the dancing continues until long after dark. One can imagine the humming and undulating movements of the dancers as they take small steps, their hands intertwined behind their backs. Fine dust rises from the plaza. There's a scent of sweating bodies, of wood smoke and goat blood. The dancing and drumming continue for four days. In the intervals between the dances, the people gather around the shrine. An elder chants a prayer to Khambageu: "Father Khambageu, bless us and open the sluices of Belwa."[7] As the chant echoes across the plaza, the trumpet sounds repeatedly, as if the god is listening to, and granting, the petitions embodied in the prayers. All agricultural work ceases during Mase. Only irrigation continues according to long-established routine.

ANOTHER EXAMPLE AMPLIFIES the picture, this time that of the Pokot, whose homeland lies nearby, to the north. The Pokot are still furrow farmers, for this simple and highly effective form of irrigation allows them to increase productivity beyond that of dry agriculture and, above

all, permits them a flexibility that was not afforded the farmers of Engaruka. Pokot farmers live in the Cherangani and Seker hills, at a higher
altitude than lowland Pokot, who are predominantly cattle herders.
Anthropologist Matthew Davies recently lived among Pokot farmers
who inhabit the north-central Cherangani Hills, especially the low-lying
Wei Wei Valley.[8] Nearly ten miles (sixteen kilometers) long and about
three miles (five kilometers) wide, the Wei Wei has steep sides and generally fertile soils. The slopes, with their spurs, known as *kuroks*, and numerous seasonal and perennial streams, provide the background landscape
for delivering water to the fields on the valley floor. The *kuroks* also form
political and social units for subdividing local Pokot society, an important consideration in an often densely irrigated valley.

Rainfall varies greatly, in both time and space, during a single rainy
season from March to August, with over forty-seven inches (120 centimeters) falling at higher elevations and much less at lower altitudes. Today,
the Pokot rely heavily on maize, sorghum, and finger millet. The latter
arrived in the area as early as 2000 B.C.E., but maize, being a New World
import, has only been grown on a large scale since sometime during the
twentieth century, with strong encouragement from former British colonial authorities. Sorghum and finger millet were the earlier staples, the
former requiring at least supplementary irrigation during dry intervals
of the rainy season and regular water during the dry months. Maize is
grown at higher, wetter elevations and does not normally involve irrigation. This may be why the Pokot have moved to higher altitudes over recent generations, a shift that has somewhat lessened their dependence
on irrigation. However, the furrow agriculture practiced by the Pokot
reflects agricultural methods that go back some distance into history,
perhaps several centuries.

Unlike many subsistence farmers elsewhere in tropical Africa, the
Pokot, like their Marakwet neighbors to the south, invest heavily in
their land and the infrastructure of their farming. Not for them the
regular shifts to uncleared land, a pattern that results in the movement
of entire villages every generation or so (and often more frequently).
The Pokot terrace their land and irrigate it with an elaborate system of
hillside and lowland furrows that take advantage of the varied, and strategically useful topography. Not that this necessarily results in a static

Figure 3.1 *A Pokot irrigation furrow taken from the Wei Wei River and running in the valley bottom through maize fields and banana plantations. The furrow is a little over 1.9 miles (3 kilometers) long, with a large stone-and-brush-built intake dam and extensive stone revetment along its upper course. (Courtesy: Matthew Davies, British Institute in Eastern Africa, Nairobi.)*

farming landscape, where the same families and households cultivate the same fields for generations. Rather, the irrigation is part and parcel of a constantly changing, highly flexible agricultural environment.

The Pokot know every foot of their irrigation channels from daily use. Matthew Davies mapped the defiles with a GPS device—about 44 miles (70 kilometers) of them. Some were only a few hundred feet long. Others extended over several miles. The longest furrow of all covered 3.1 miles (4.9 kilometers), stitched together from two branches. Some now-disused channels may have been even longer. Like all furrow systems, the Pokot channels are gravity driven, the steepest gradient being about one in three (just under 30 percent), the shallowest a mere one in

Figure 3.2 *Another Pokot furrow, taken from the Tororro stream, in the Wei Wei Valley. This furrow contours across the valley sides for some 2.5 miles (4 kilometers), supplying water to numerous maize fields. The furrow is substantially built, especially in its upper course, where large stone slabs reinforce the downslope side of the furrow. (Courtesy: Matthew Davies, British Institute in Eastern Africa, Nairobi)*

seventy to eighty (less than 1.5 percent), an impressive feat of careful engineering.

Hill furrows run off the seasonal streams that descend the valley slopes. The intake is a small mud-and-brush dam that raises the water level and allows gravity to divert water into the waiting furrow. Now the channel is contoured out of the stream gully with considerable ingenuity. The builders carefully revet the watercourse close to the intake. Then they lead the water along a twisting route, around and over rock faces, avoiding large boulders. Occasionally, they use a hollow log as a simple aqueduct. As the gradient eases, construction becomes easier.

Now the furrow is a small, gravel-lined defile cut into the hillside, sometimes reinforced with an artificial embankment. A furrow is no more than 2.6 feet (0.8 meters) wide and 1.5 feet (0.5 meters) deep, sometimes twisting and turning around as many as three *kuroks*. At the end, it cascades down the crest of a spur, which allows the furrow to serve farmers on either side of the spur as well as maximizing the amount of irrigated land. Generally, the land on either side belongs to no more than two clan sections, which makes for relatively straightforward management.

The furrows that lead from the major rivers on the valley bottoms are wider than those on the hillsides. Large brush, mud, and stone dams divert water into the channels, which are normally stone-lined, especially close to the intakes. They are longer as well as wider, and sometimes as much as 3.3 feet (1 meter) deep, with much shallower gradients that promote sediment buildup. In the long term, the effect is like that of river levees, the banks climbing above the level of the surrounding terrain as the owners of the furrows deepen them. Valley furrows are much easier for small groups to construct and maintain, which makes for considerable flexibility in terms of ebb-and-flow irrigation farming.

Earthen feeder canals lead from both hillside and valley channels into the surrounding fields. Their owners simply plug or unplug the intake from the main channel with brush and stones as needed. A series of smaller feeder channels disperses the water to individual fields. The water runs through a small channel above each plot. All the farmer has to do is breach the low embankment in several places. He then disperses the water across the plot with his hoe. Usually, the major feeder canals run downslope through a series of low, terraced fields, where the smaller channels disperse the water along the terrace lines.

Given that much of their agriculture depends on rainfall rather than water transport, the Pokot appear to place less importance on irrigation than the Engaruka farmers once did. Hill furrows, with their associations with *kuroks* and involvement with one or two clan sections, are a relatively straightforward management problem. Valley furrows are another matter, for they flow through the lands of numerous clans. The Pokot administer both hill and valley furrows through institutions known as *kukwa*, councils of circumcised men. The *kukwa* meets irregularly, its membership drawn from fellow kin and village members. The members vary in number, but a *kukwa* concerned with irrigation comprises the users of the furrow it administers.

Hillside *kukwas* tend to encompass both irrigation beneficiaries and political units, which means that members know one another well, having no illusions about their abilities and individual needs. This makes for relatively straightforward decision making. Maintenance parties are quickly organized, and members exchange information about crops they intend to plant and their status during the growing season. Irrigation water comes in units of a day, distributed on a rotating basis. At least, this is the ideal, but the negotiations continue, month in, month out, as individual needs change. All of this makes for a flexible allocation system, which works well, since fairly small numbers of people are involved.

Valley furrows involve more complex administration, in the hands of irrigation *kukwas*, not politically based ones. A much more diverse group assembles, few of its members related by ties of kin. General decision making about allocations is more constrained, with each block of land owned by a village receiving an equal allocation, the water then being distributed after discussion between individual users and *kukwa* members, who do know one another and often share ties of kin. Everything depends on the knowledge and gravitas of respected elders, as well as long-established rules of etiquette and precedent passed from one generation to the next. Davies calls this "a subtle form of oligarchy," an apt description of a formalized political process that leads to efficient communal decision making. The *kukwa* functions because of these established rules, and on the assumption that peer pressure from fellow kin will ensure participation in communal maintenance and proper use of water allocations.

Again, this is the ideal. In theory, everyone receives equal treatment and the same allocations. In reality, other factors come into play. A skilled orator at *kukwa* gatherings may argue a persuasive case and receive more water. Strategic gifts of beer, milk, goats, or sheep for feasts and ceremonial occasions can bring more water. Sometimes, a farmer will "purchase" more water by contributing labor to neighbors, or by bartering some of a current allocation for part of a future one. The system is in a state of constant flux and individual negotiation, especially when there is plenty of water and communal needs are low.

Pokot irrigation works well because of this flexibility, and also because the maintenance required to keep the furrows in good working order is much smaller than one might think. Twice a year or so, *kukwa* members work on repairing the intakes and clearing the main channel. Individual sections assume responsibility for bank repair and fixing breaches that may occur in heavy storms. The *kukwa* appoints someone who lives close to the intake to monitor it on a daily basis, diverting water away during storms and maintaining the critical early reaches of the channel. This "furrow-man" receives gifts of beer and goats during important feasts.

It would be a mistake to think of Pokot irrigation as a static, self-sustaining entity. It has endured for many generations, but its durability comes from its very flexibility, its ability to expand and contract in the face of social changes, different forms of government, tribal warfare, and climatic shifts. Engaruka reached the limits of its irrigable land and available water supplies. Neither the Marakwet nor the Pokot have ever faced the specter of starvation triggered by a rising population, a shortage of agricultural land, or a shortage of water. The Pokot have responded to change by shifting their villages to adjacent land or higher elevations. They have never confronted a totally degraded, effectively waterless landscape like that of Engaruka, whose people had to move away.

THE MARAKWET SAYING about irrigation channels and wives is a wonderful way to define their relationship to water. Their furrows are simple but carefully constructed, never-static structures, modified constantly as the farmers expand a patchwork of fields, abandon sorghum plots, or clear new land to accommodate rising numbers. But the cultural equation is far more complex than this. The hydrological solutions may be relatively straightforward, but they are based on hard-won experience and the subtle crosscurrents of public opinion. This is one of the reasons why reverence for the ancestors is of such great importance in subsistence-farming societies everywhere. Those who have come before are the symbols of the unfolding rhythms of the fields—planting and birth, growth and harvest, the fallow months of the off-season that symbolize death. Their experience, the complex, ever-shifting realities of kinship, carefully arranged marriages, conflicts over inherited allocation rights—these and many other subtle cultural traditions, values, and institutions form a complex partnership with the seemingly straightforward task of delivering water to near and distant fields.

The Marakwet and Pokot systems are just two examples of the small-scale irrigation that still flourishes among subsistence farmers in many parts of the world. But what happened when the scale of water management and delivery expanded exponentially, when irrigation-dependent farmers found themselves packed into landscapes where self-sustainability depended on elaborate canal networks and equitable water delivery to thousands of people? The Hohokam farmers of the extremely arid and hot Sonoran Desert and the water temples of Bali, a land of abundant monsoon rainfall, provide two fascinating and contrasting examples.

CHAPTER 4

Hohokam: "Something That Is All Gone"

OCTOBER 2008. The statistics bubbled in my mind as our propjet trundled along above the Arizona desert: Over one million eight hundred people in Phoenix itself, a further twenty-two cities surrounding it in the Valley of the Sun, forming the largest metropolitan area in the United States; about 7 inches (17.8 centimeters) a year in rainfall; average summer temperatures over 100 degrees Fahrenheit (38 degrees Celsius) for three months a year, with occasional peaks as high as 120 degrees Fahrenheit (49 degrees Celsius). I gazed out at the arid reaches of the northwestern Sonoran Desert, not a tree in sight, only dry watercourses zigzagging across sandy wastes. After a while, I dozed off, mesmerized by the monotonous engine noise and the endless dry landscape.

The engine pitch slowed. I woke abruptly as we descended from wilderness into an almost surreal urban landscape. The sun was setting behind us, casting long shadows and softening the glare. But Phoenix shimmered beneath us; broad streets led into the far distance, a palimpsest of huddled apartment buildings and carefully ordered subdivisions scattered over arid terrain, the margins stopping suddenly in undeveloped sand. High-rise office buildings towered incongruously over the desert. Everything was yellow or buff colored except for occasional splashes of green—golf courses and parks, irrigated farms that seemed to have sprung without notice from the dry landscape. Dozens of blue swimming pools adorned suburban yards. Over to the left, the long, straight line of an aqueduct ran to the distant horizon. I remembered that all of this urban sprawl depended on finite water supplies pumped from deep beneath the earth or delivered from afar. Coming from a California in the midst of a multiyear drought, I idly wondered just how long it would be before promiscuously expanding Phoenix imploded in the face of chronic water shortages triggered in part by a culture of urban excess and waste.

We descended gradually, bumping gently in the late-afternoon turbulence, and passed over the meandering Salt River, its waters yellow-brown in the hazy sunlight. Then I recalled the Huhugam, "something that is all gone," ancient people known to archaeologists as the Hohokam, who had farmed this inhospitable desert for more than a thousand years.

IN 1882, SMITHSONIAN anthropologist Frank Hamilton Cushing entered the Salt Valley in the northeastern reaches of the sweltering Sonoran Desert, in what is now Arizona. He climbed to the top of an earthen monument and was astonished to find himself in the heart of a long-abandoned Indian settlement. He wrote, "It was one of the most extensive ancient settlements we had yet seen. Before us, to the north, east, and south, a long series of . . . house mounds lay stretched out in seemingly endless succession."[1]

Cushing was by no means the first outsider to explore the valley. A small town rose near the Salt River in 1860. Five years later, the United States Cavalry established Camp McDowell in what is now Maricopa County, centered around Phoenix. Enterprising visitors to the camp observed not only the nearby eroded mounds of what were once adobe structures, but also the remains of extensive irrigation canals that had once brought water from the Salt River to wide tracts of now-abandoned maize fields. In 1867, Jack Swilling, a former soldier and jack-of-all-trades, established the Swilling Irrigating and Canal Company. The company flourished, for its canals often followed those of the ancient farmers. Farming prospered, despite the extremely hot summers. On October 20, 1871, a mass meeting of settlers appointed a committee to select a town site. After considerable debate, committee member Darrell Duppa proposed the name Phoenix, for a city rising phoenixlike on the ruins of an ancient civilization.[2]

Phoenix and its surrounding communities have paved over much of the Hohokam universe. But the long-vanished farmers reappear with persistent frequency, under the foundations of modern buildings razed for new development, in the pathways of expanding interstates, even in backyard gardens. For the most part, the traces of their presence are inconspicuous, requiring careful dissection with spade and trowel. Only a few more notable adobe structures still lie aboveground. Today's inexorable urban sprawl makes it hard to believe that the Salt Valley was the most populous and agriculturally productive valley in the Southwest before 1500 C.E. The land looks barren and utterly dry, yet it has fertile soils and lies near major drainages. Between 450 and 1500 C.E., the Hohokam living near the perennial Salt River adapted brilliantly to this seemingly desolate environment, refining their agriculture and water management from one generation to the next. Over more than ten centuries, they built vast canal networks up to twenty-two miles (thirty-five kilometers) long and irrigated large tracts of arid land up to seventy thousand acres (around twenty-eight thousand hectares) in extent, all of this without the elaborate panoply of state government or highly centralized management.[3]

Archaeologists identify the Hohokam by their buff to brown potsherds, which abound in the basin floors of southern Arizona. If we use such vessels as a criterion, then we can trace the extent of the Hohokam over more than thirty thousand square miles (seventy-eight thousand square kilometers) of southern Arizona—an area larger than South Carolina. How meaningful the term Hohokam is in terms of cultural identity is uncertain, for there was great variation from one community to another. In general terms, Hohokam groups shared a common ingenuity as farmers, a superb ability at irrigation agriculture, and an architecture of adobe dwellings. There were none of the elaborate, multistory pueblos of Chaco Canyon and Mesa Verde here, but rather, a distinctive ceremonial architecture based on adobe platform mounds and ball courts, the latter apparently an import from Mexico. (Platform mounds served as foundations for elite residences and rituals.) There were common understandings and, above all, profound values of reciprocity, of mutual obligation.

It cannot have been any other way in a world where rainfall was at best sporadic and usually highly localized. This was a desert landscape

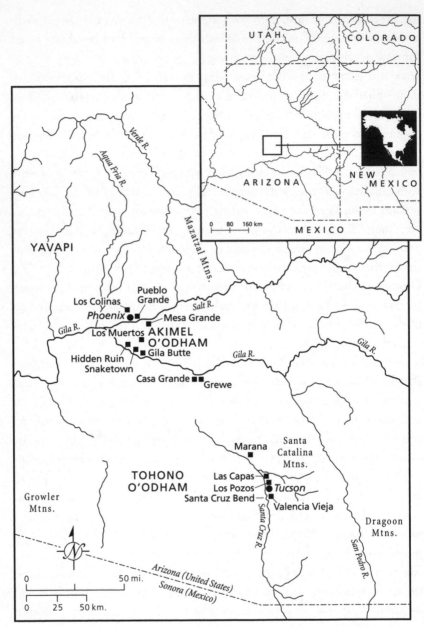

Figure 4.1 *Map showing locations in chapter 4.*

inhabited by people in a constant state of social and political flux, reflected in the great ethnic diversity of the region's Native American populations today. For all this instability, everyone must have paid close attention to kin living near and far, for in a world where everything depended on careful water management, you were to a great extent dependent on the labor of your kinfolk and others. As far as we can tell, for most of the Hohokam's tenure in Arizona there were no great leaders who supervised water harvesting, canal construction, or allocation of water to the fields. Everything functioned at the village level, through the same ancient mechanisms of family and kin as we have seen in east Africa's Rift Valley.

A SUMMER DAWN unfolds a thousand years ago, the first flecks of gray on the eastern horizon. The people are already hard at work in the cool of morning. Rows of men with wooden digging sticks and stone-bladed hoes dig into the sandy soil along a canal line drawn carefully from the riverbank into the desert. Women and children line up with baskets and heft the loose sand up a short slope to form earthen embankments on either side. As the men dig, clad only in loincloths, they chant and sing to give rhythm to the backbreaking labor. The sun rises, and sweat pours down their backs. They pause to drink water from cool gourds, knowing full well that thirst is deadly. The day will be too hot for prolonged work, so they'll come back after the midday heat to dig . . .

Hohokam canals flow outward from the Salt River like the tentacles of a giant octopus. They bifurcate and bifurcate again, once full of gently flowing water transported for mile after mile by the forces of gravity. Omar Turney, a city engineer for Phoenix, mapped the ancient Hohokam irrigation systems during the 1920s. He walked over the river basin and consulted old maps and historical records to reveal what he called "the largest single body of land irrigated in prehistoric times in North . . . America."[4] The lower Salt Valley, where downtown Phoenix now stands, supported mile after mile of irrigated fields and dozens of farming communities. The scale of the irrigation works boggles the imagination. In the downtown area alone, 300 miles (483 kilometers) of canals formed fourteen irrigation networks that watered 400 square miles (around 7,700 hectares) of fertile river-basin soils. The Gila River to the south, with its four irrigation networks, watered nearly 19,000 acres (7,690 hectares) of closely packed fields a thousand years ago. In the heart of this carefully engineered landscape, on the second terrace of the Gila, stood the 250-acre (101-hectare) Snaketown site, with its ceremonial ball court. Six miles (10 kilometers) of irrigated land and smaller settlements lay upstream and downstream. The dense cultivation extended as much as 2 miles (just over 3 kilometers) from the riverbanks.

The Gila and Salt rivers received their water from highland watersheds. The river flow varied from season to season and from year to year, but provided generally reliable water supplies for the Hohokam's fields. Away from the great rivers, farmers relied on both summer and winter rainfall. They would trap the occasional tributary flood, use terraces and small dams to trap other water, and place their crops with meticulous care to avoid extremes of heat or cold. There was nothing the Hohokam did not know about farming in arid landscapes, which they supplemented with hunting and plant foods.

The canal systems began at the river, where a weir raised the water level and directed the water into the canal. A head gate regulated the amount of water passing into the system. From there, the water flowed along large distribution canals that were up to eighty-five feet (twenty-six meters) across and twenty feet (six meters) deep. The size of the canal diminished away from the river, a technique that ensured an even water flow. A steady flow ensured that too rapid a current did not erode the earthen sides. If, however, the current was too slow, it would deposit silt and clog the defile. Control gates lay at intervals along the main canals. When closed, they caused the current to back up, creating a head of water and allowing the farmers to regulate the flow down-canal. Feeder channels carried water through wicker and stone gates along long branches off the main canals, in turn leading to much smaller defiles that fed grid-like field systems, each with its own water supplies constrained within banks like a crowded chessboard.

The amount of communal labor required to construct and maintain these irrigation works was enormous. Reconstructions of ancient canals

Figure 4.2 *Southwestern archaeologist Emil Haury standing in an excavated Hohokam irrigation canal. (Courtesy: Arizona State Museum, Tucson)*

suggest that as many as 28.25 million cubic feet (800,000 cubic meters) may have been excavated to construct one major canal system alone. If a single worker removed 106 cubic feet (3 cubic meters) of soil a day, it must have taken more than twenty-five thousand person-days to build many of the canals, a process that must have taken years.

The Hohokam's irrigation systems produced bountiful crops and often supported multiple harvests year after year. We know of their success from the long duration of their canal networks, which began on a small scale but expanded and remained in use, as tree rings and radiocarbon dates tell us, for a thousand years. But to construct and maintain the canals, weirs, and gates, and to allocate water and resolve conflicts between communities and individuals, required a complex social and political structure that developed over many centuries.

HOHOKAM CULTURAL ROOTS go far back into the past, but the group's origins remain somewhat of a mystery, except for a general impression that their farming began further south in the desert.[5] Like Phoenix, present-day Tucson to the south lies on much earlier human foundations. As early as 1500 B.C.E., and perhaps earlier, farmers were cultivating small fields alongside the Santa Cruz River. Many of their plots lay along floodplains where tributaries joined the river, or where subsurface boulders or impermeable sediments caused groundwater to back up closer to the surface. Like early cultivators in the Near East and elsewhere, these maize agriculturalists were expert at selecting the best soils for their crops, which they irrigated with short gravity-fed canals dug into the floodplain. The result was a checkerboard of small, irrigated plots carefully laid out in close juxtaposition to make maximal use of both soil and water.

Such farming continued for about a thousand years. This is hardly surprising, as people had been cultivating maize in the region since at least 2000 B.C.E., when this staple crop passed northward into the Southwest from northern Mexico. An expertise in arid-lands farming with careful water management was already well established. Like agriculture itself, irrigation was not a dramatic invention. Every farmer knew that water flowed downslope and that it could be diverted with some digging and careful use of gravity. Irrigation in the Santa Cruz Valley was a simple extension of long-ago-acquired farming expertise in a demanding, drought-prone environment.

This basic form of irrigation was transitory, in the sense that the farmers engaged in it dwelled in at best semipermanent villages. They lived in circular wicker shelters, staying close to their crops during the summer growing season, then moving onto higher ground and across the surrounding landscape in search of game and wild plant foods during the rest of the year. For many months, they followed the ancient hunting and gathering practices of much earlier societies. There was none of the great elaboration of irrigation agriculture that was to follow in later centuries.

The Santa Cruz settlements are nothing much to look at, their excavation revealing numerous small, circular hut depressions surrounded by storage pits cut into the alluvium. The effect is just like impressions

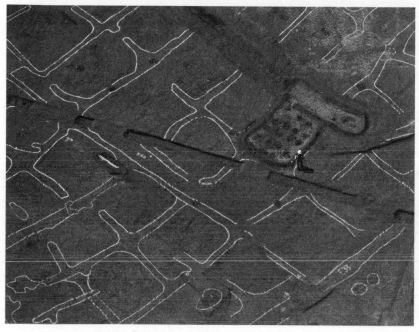

Figure 4.3 *Small pithouses used by early irrigators along the Santa Cruz River. Storage pits can be seen on many of the circular floors. (Courtesy: Henry D. Wallace)*

made in flour dough. It is as if "someone went wild with a set of cookie cutters," writes archaeologist Henry Wallace, who has unrivaled experience with these pithouse sites.[6]

About the time of Christ, the short-lived hamlets dropped out of sight. We have no idea why, but the Santa Cruz and other southern Arizona rivers were downcutting their beds sharply into deep channels. The farmers' irrigation canals now stood high above the river flow and were bone-dry. This may have been when many ancestors of the Hohokam moved northward into the Salt and Gila valleys, where large-scale irrigation began in about 450 C.E.

INEVITABLY, A NEW, but still little-known, society emerged from much earlier foundations, one where the control and organization of irrigation works was a priority. A village at Valencia Vieja, near the Santa Cruz River south of Tucson, tells part of the story. A cluster of five to ten houses rose here in about 425 C.E., a hamlet typical of dozens of others in the region, probably in a hitherto-unoccupied area, where the people had to dig new irrigation canals. In about 500, Valencia Vieja suddenly doubled in size, thanks to an influx of newcomers.[7] The hamlet became a village; a plaza flanked by at least eight larger adobe structures rose around the open space. The excavators, Henry Wallace and Michael Lindeman, believe that these housed an important lineage or clan leaders who now lived permanently at the village, with the houses of kinfolk behind them. A group of kin leaders, more permanent settlement, more mouths to feed—this may have been a settlement ruled by a village council.

Wallace and Lindeman think that two forces brought the villagers together. They needed to dig increasingly large irrigation works ever further from the river. This would have required large numbers of diggers to work on a communal enterprise that benefited everyone. Then there was the issue of who controlled the land and irrigated water at a time when the local population was growing rapidly. By now, people were developing large field systems that effectively locked them into the landscape. For the first time, landownership and water rights created a sense of collective identity and a social order that revolved around the common interests: water and land. This same consolidation, the appearance of permanent villages tied to field systems and irrigation canals, appeared at Snaketown, to the north, at about the same time.

The plazas at Valencia Vieja and elsewhere proclaimed a reorganized society. For the first time, people lived in the same communities for centuries, where they acquired collective identities based firmly on their kin affiliations. Each plaza became a public arena, where leaders, some of them perhaps shamans, performed and served as symbolic links to the supernatural realm, where the ancestors dwelled. Burials appear in these open spaces, as if the ancestors were buried in the heart of the community, in places where rituals bound together the living, the dead, and the supernatural forces that controlled human existence.

Hohokam culture as archaeologists know it appeared around 450 C.E., at the moment when plazas first came into existence. New pottery forms also came into being, and improvements in grinding technology made food preparation somewhat easier. More diverse cooking vessels now allowed mothers to wean their children earlier and feed them on soft foods. At this juncture, too, new varieties of maize appeared in Hohokam fields. All these changes fueled population growth, at a time when much ritual activity still focused on the irrigation cycles, the times of planting and harvest, and ceremonies that helped bring summer rain. As we shall see below, the heart of Hohokam beliefs was the notion of bringing rain to water the crops, fill irrigation canals, and encourage the growth of edible wild plants. Unfortunately, we know nothing of the rituals that accompanied the irrigation cycles or the planting and harvesting of crops. We do know, however, that water rituals, described below, had a central place in ceremonial life.

Nor can the modern-day O'odham people, who followed the Hohokam in the region, throw any light on irrigation rituals. They believe that the Hohokam (Huhugam) were their predecessors. Daniel Lopez, a Tohono O'odham elder, writes, "We just say that we go back to the Huhugam. We are here today, but we know that some time in the future we will be called the Huhugam." There are still landmarks and sacred places that remind the O'odham of the Hohokam, among them a cave where I'itoi, the Creator, dwelled when he was on earth. Whenever people were in danger, he came to help, as well as destroying dreadful monsters like the Eagle Man and Ho'oki Oks. Casa Grande, or Siwan Wa'a Ki, the largest site, figures in O'odham legend, and people still go there to pray and sing songs to Huhugam spirits. "As we breathe the holy air that gives us life, we can feel the power of the ancestors. When we see the stars at night and hear the owl, some of us feel strongly that we are part of the ancient past."[8]

Most Hohokam families lived in small, single-room houses built of pole and brush and covered with hardened clay. Groups of dwellings lay around small courtyards, suggesting that extended families lived close to one another. The size of Hohokam courtyard groups varied considerably: One at the Grewe site, on the Gila River, had as many as twenty-four houses, covering sixty-five hundred square feet (six hundred square meters).[9] Courtyard groups and the irrigated fields and gardens along the canals associated with them passed from one generation to the next. Thus, the household, in the wider sense of an extended family, was the primary mechanism for both controlling landownership and the use of irrigation water over long periods of time.

Phoenix's Sky Harbor Airport lies on the Salt River floodplain. When archaeologists surveyed the land under the runways and airport buildings, they found extensive tracts of once-irrigated land, as well as large and small houses built alongside the fields. Here, people camped during the growing season, sleeping and cooking in the larger structures, which had hearths, and using the smaller ones for storage and other purposes. These were casual, temporary structures, rebuilt again and again over the centuries. But this very rebuilding confirms that the same households farmed the same plots of land over many generations.

Early permanent villages founded along the Gila and Salt rivers coincided with the construction of the first large-scale canal system. If the O'odham are any guide, those who participated in canal construction had the first pick of the best land. Later arrivals would have occupied less desirable acreage. Over time, the first comers, who enjoyed the best access to prime irrigated land, would have acquired wealth and status,

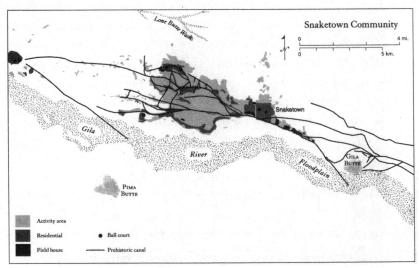

Figure 4.4 *Plan of the Hohokam irrigation systems around Snaketown, on the Gila River. (Molly O'Holloran, School for Advanced Research, Santa Fe, New Mexico)*

in part because of their more productive fields. In later times, Hohokam communities varied greatly in size. Some, like those on the Gila River, south of Phoenix, ranged along single canals that lay parallel to the river on either bank. Such communities may have covered fifteen square miles (thirty-nine square kilometers) and contained as many as 2,550 irrigated acres (1,031 hectares). Salt River communities lived amid branching canal networks that traversed the Phoenix area. Suzanne Fish, who has studied Phoenix Basin communities of 1100 C.E. and later, estimates that community territories averaged fifteen square miles, a very similar figure to that from the Gila Valley.[10]

WHO, THEN, CONTROLLED the distribution of water through these ever-growing irrigation systems in a society where reciprocity, sharing, and community were all-important to survival? At the Grewe site, the excavators looked for the largest houses, which they assumed were those of the wealthiest families. They found that these clustered in a few courtyard groups and were occupied much longer than smaller dwellings. Perhaps these particular households were able to maintain their advantageous position in local society over several generations. But were the people who lived in such households authoritarian leaders, who competed for power and prestige and acted as village leaders? Were they the individuals who transformed Hohokam life over many centuries? Or did their influence never extend beyond domestic concerns?

After excavating Grewe, Douglas Craig and Kathleen Henderson believe that households were key players in the power equation.[11] They discovered a communal cooking area with dozens of pit ovens cut into the soil. This lay on one side of a crowded residential area, where hundreds of dwellings stood close together in courtyard groups. On the other side was a public plaza with a ball court. Craig and Henderson think that the nearest groups, which included the largest and wealthiest, controlled activities in the cooking area and sponsored the feasts that were prepared there, a way of acquiring prestige and social status. At the same time, the events reaffirmed property rights held by the richest households as well as fostering a sense of communal identity. There are signs, too, that the same households also subsidized craft production, the manufacture of items that were traded to other groups in the region. These would have included cotton textiles, finely made pottery, shell jewelry, and stone tools. Judging from the distribution of craft tools, finished tools, and manufacturing detritus, the artisans responsible for their making lived in small courtyard groups on the margins of those of the wealthy.

Interestingly, people living in courtyard groups outside the Phoenix Basin appear to have moved more frequently. In the Tucson Basin to the south, Hohokam households irrigated their fields from the Santa Cruz River. However, the river channel shifted constantly, the floodplain changed character, and the farmers were constantly adjusting to new hydrological circumstances. In contrast, the Gila River floodplain remained relatively stable throughout the six-hundred-year life of the Grewe settlement. This meant that each household could reasonably assume that it would cultivate the same irrigated fields year after year. This stability contrasts with the experience of O'odham farmers along the Gila during the nineteenth century. They used the same irrigation methods as the Hohokam, but had to move their villages much more often.

The household lay at the core of Hohokam irrigation works and of society as a whole. Its members helped build and maintain a sophisticated canal network. They provided the intensive labor needed to sustain an intensive agriculture regimen for nearly a thousand years. How did this system work, with all the close cooperation that it required? Judging from historic practice among the O'odham, individual households held claims to plots of land that passed from one generation to the next. However, just as in east Africa, rights to irrigation water belonged to the community as a whole, the water being allocated to each household according to the amount of land it cultivated. Households shared water rights. Thus, they had an interest in protecting these rights and protecting their investment in the canal works that brought water to their fields.

Craig and Henderson point out that there must have been a dynamic tension between the individual interests of the households and the collective ones. The two tension points were always in play, balancing one another out over years, generations, and centuries. Neither side won, but the resulting balance enhanced the long-term stability of a society that depended heavily on canals to bring water to soils that were otherwise useless.

On this wider scale, we can think of Hohokam society as a series of what the archaeologist David Doyel once called "irrigation communities." He imagined Hohokam irrigation in terms of canal networks branching from a single river intake. These networks connected an array of interdependent villages, whose households shared the labor of canal construction, maintenance, and management in a peaceful manner. Doyel identified at least six such irrigation communities along the longest Phoenix Basin canal network.[12] By 1100 C.E., these larger agglomerations each comprised a prominent village with communal structures such as a ball court, a plaza, or a platform mound and outlying smaller settlements and farms, the whole surrounded by carefully laid-out and intensely cultivated farmland. A web of secular and ritual relationships connected every individual in Hohokam society with a wider world.

RITUAL PLAYS A central part in all the human societies we visit in these pages, for it is the glue that holds society together. The Hohokam were no exception. Each season had its rituals, its commemorations of impending rain, planting, the harvest, the ancestors, and the powerful forces that presided over the fate of humanity. Ritual provided explanations of the cosmos, of the realities of the world, of human existence. And in a society that depended on communal labor, ritual provided the critical links between those who presided over ceremony and ritual observance and those who labored on the communal irrigation works that sustained human life. Ritual provided the context for the latter endeavors, reflected in two remarkable architectural features that were the backdrops of later Hohokam society: ball courts and platform mounds.

One form of Hohokam settlement centered on what another archaeologist, David Wilcox, called "ball court communities."[13] Wilcox observed that ball courts were spaced at relatively even intervals along Hohokam canal networks. These basinlike structures were arenas where people from the surrounding countryside would gather for feasts, trading activities, and all kinds of social interaction, as well as ball games between different villages. We can even go so far as to delineate, at least provisionally, the boundaries between different ball court communities along the Phoenix Basin canals, but it's interesting that ball courts also occur in other areas of the Hohokam world, where irrigation works were on a much smaller scale. Some of them even stand in places where there was no irrigation whatsoever. This suggests that Hohokam communities far outside irrigated areas also shared common beliefs and ideals, a need for rituals that reinforced the need for cooperation with others.

Ball courts and the ball game associated with them are an ancient tradition among Central American civilizations that endures in modern times. In the game's most flamboyant form, the players endeavored to shoot a rubber ball through a stone hoop high above the ground with their hips, without using their hands or legs. Elaborate ceremonials and vigorous betting accompanied Maya and Aztec ball games. It is said that the losers were sometimes sacrificed to the gods. There's no reason to believe that Hohokam ball games involved such high stakes and achieved such elaboration, but ball courts are a pervasive feature of the larger communities. We know of more than two hundred of them, large and small, shallow depressions with plastered or stamped earthen floors, surrounded by shallow banks where up to seven hundred spectators could witness the games.

We can imagine the crowd gathering in the cool of late afternoon, standing cheek by jowl, in places overflowing onto the earthen floor. Young men jostle one another, pushing for a better view. Dogs scurry; children cry; the scent of wood smoke drifts through the still-warm air. Below the teeming watchers, the brightly painted players from neighboring communities careen into one another, keeping the rubber ball in play with their hips. The sweating crowd shouts and chants, drums are beaten and rattles sound as the setting sun bathes the court in deep shadows.

Major ball games were special occasions, sometimes attracting people from miles around. The largest courts were up to 250 feet (76 meters) long and 90 feet (27 meters) wide, dug up to 9 feet (nearly 3 meters) into the subsoil.[14] Ball courts are oriented in various directions, as if different courts were used to commemorate specific events during the year.

Hohokam communities built ball courts between about 700 and 1100 C.E., when they abruptly gave way to entirely new ritual structures: platform mounds.

Quite what form the ball game itself took remains a mystery, but there is no question that it originated in Mexico, where commoners played a version of the contest that required each side to cast a rubber ball back and forth without its touching the ground. Three such balls have been found at Southwestern sites. Judging from historical analogies, the contests were the culmination of days of feasting, trading, and social interaction that enhanced a sense of communal identity.

No Hohokam canal system could operate without input from dozens of small communities, the farmers who used the defiles to water their fields. As far as we can tell, the administration and planning of earlier irrigation works lay in the hands of villagers and their kin leaders, with planning, construction, maintenance, and water allocation a matter of carefully worked-out consensus, just as it is in east Africa to this day. But the appearance of ball courts hints at new styles of leadership and a greater emphasis on communal rituals, even social inequality, where the kin leaders of earlier times now became more authoritarian, perhaps a function of higher population densities and the need for more-closely-organized water supplies. Ball courts may have been part of this process, a form of ritual observance that commemorated the passage of the seasons, the coming of river floods, and new growing seasons. After 1100 C.E., the courts gave way to platform mounds, elevated structures filled with earth to form a higher place, an ancient form of Hohokam architecture that now assumed much greater importance. Between 1250 and 1350, platform mounds grew dramatically in size, standing higher than a person and composed of thousands of cubic feet of fill that involved communal labor on an impressive scale. More than 120 platform mounds have been found at over ninety-five sites, most of them in the Phoenix Basin. Some are over 12 feet (3.5 meters) high, built within an adobe compound, with as many as thirty rooms on the summit. These were enclosures of limited access, unlike the ball courts of earlier times. It is as if Hohokam society became more hierarchical, with only a few individuals having access to the precincts within the enclosures. That these people controlled and managed water supplies seems unquestionable.

Part of this management appears to have involved public rituals, perhaps an elaboration of earlier ball court ceremonials.

Everyone agrees that the platform mounds were places for ritual performances. But did a small number of privileged people or families live there, separated from most of the population? Certainly, the mounds were built so that the activities atop them could be seen by large numbers of people. Public performance was an integral part of fostering a common identity, of linking leaders with the general population, of providing the context for the arduous communal labor that brought water to the fields. The mounds were also trade centers and the places where leaders would distribute food and other commodities, and where outsiders were greeted. The dynamics of water management were now more competitive and were vested in social status in ways unimaginable even a few centuries earlier. Territorial markers and displays of prestige and power—the larger the platform mound, the more labor required to build it, and the greater the status of the individuals or group responsible for its construction. Perhaps as many as fifty thousand person-hours went into erecting the mounds and mound compounds at Pueblo Grande and Mesa Grande, the largest mounds in the Phoenix Basin. By this time, Hohokam society had developed an elaboration unknown in earlier times, with the emergence of powerful leaders, who inherited large tracts of land, controlled elaborate canal systems, and commanded the loyalty of hundreds, if not thousands, of people.

The platform mounds may mark another significant change in Hohokam society.[15] Ball courts were open depressions, accessible to large numbers of people. Platform mounds reached for the sky, stood out on the landscape, and were accessible to only a few. The shift in ritual architecture may have coincided with a change in the Hohokam's relationship with the supernatural world. Ball court construction peaked about a century before platform mounds became commonplace, at a time of widespread drought throughout the American West, documented in tree rings. It was as if a few members of society elevated themselves in both material and spiritual terms above everyone else, whereas in earlier times the relationship between the living and the ancestors, with the underworld where humans originated, had been more important. Now, perhaps, the close spiritual relationships were between a few individuals

Figure 4.5 *Reconstruction of a Hohokam platform mound and residential compounds at the Marana Mound site. (Courtesy: Pamela Kay)*

with unusual powers and the water deities of the supernatural realm. We will never know.

WE HAVE NO means of fully deciphering the complex ritual beliefs of Hohokam communities. However, we can reasonably surmise that many of their beliefs revolved around rain and water, around bringing water to nourish growing crops. Unlike most Native American groups, early Hohokam culture had many features that recall Mexican societies, among them pithouses, pottery designs, and figurines. Elite goods such as copper bells, mirrors, and tropical bird feathers arrived from western Mexico, as did architecture such as the ball court. Why not, then, ritual beliefs, such as those of the rain cults associated with the Central American rain-storm-earth god Tlaloc? Ancient Mexicans believed that Tlaloc dwelled in the mountains, from whence rain came, the origin place of streams, rivers, and lakes, the realm of lightning and thunderclouds. Mountains were water containers, as well as passages to great caves where maize was born. The metaphor carried over to flat-topped pyramids and mounds with temples atop them—conduits to the underworld. It may well be that the Hohokam adopted part of the symbolism of the ancient Mexican rain god.

There were ancient, symbolic connections between water and ceramic vessels, for, after all, mixing earth and water forms clay.[16] This symbolism extends to the decoration that adorns Hohokam vessels: water birds, fish, snakes, turtles, and other creatures that live near water. Even geometric motifs such as scrolls representing waves or serpents symbolized the connections between the realm of the living and the underworld. Stephanie Whittlesey, who studies Hohokam pots and the ritual behind them, points out that the plain vessels glitter in the sun, as if depicting the sparkle of the sun on water. The glitter comes from the schist of Gila Butte, a double-crested peak that towers 500 feet (150 meters) above the Gila River. The river curves around the butte, and Snaketown lies only a few miles west, a focal point of the Hohokam world for many centuries. The main irrigation canal for Snaketown lay at the foot of Gila Butte, where the hard bedrock raised the water table, creating the impression that life-giving water flowed from the mountain. It may be no coincidence that Snaketown was a major center of pottery manufacturing. Every vessel that left its workshops contained some of the schist from Gila Butte. Whittlesey writes, "We can imagine that rituals performed with such containers were enhanced, prayers said over them were more effective, and the stored food and water the vessels contained were considered blessed."[17]

The links between artifacts, ball courts, and platform mounds, as well as farming and ritual, were deeply embedded in Hohokam life. A strong and enduring society combined the sacred and the secular in a lasting partnership based on an ancient ideology, in which everything had multiple meanings and contributed to the long-term survival of a people for whom water defined life in unique ways.

BETWEEN 1150 AND 1450, after almost a millennium of relative stability, profound changes occurred in Hohokam society. These three centuries

were a time of unstable environmental conditions. The instability coincided in part with the major droughts of the Medieval Warm Period,
which descended on the Southwest and contributed to the abandonment of Chaco Canyon and its great houses and, later, large pueblos
at Mesa Verde and in the nearby Moctezuma Valley. In the desert, this
same period may have been a time of flash floods that altered river channels and undermined canal systems. Sometimes also, prolonged droughts
caused lengthy water shortages. We know little of these environmental
perturbations, but we do know that Snaketown lost population. Meanwhile, other communities, such as Casa Grande, consolidated formerly
separate canal systems into a single irrigation network based on one
twenty-one-mile (thirty-four-kilometer) canal that watered fifteen thousand acres (six thousand hectares) of fields. Five villages, each with its
own platform mound, lay along the canal. Another irrigation community, in present day Mesa, had twenty-one miles (thirty-four kilometers)
of canals that watered fourteen thousand acres (fifty-six hundred hectares) of fields. By any standards, this was irrigation on a large scale,
with the largest Phoenix Basin communities each supporting between
six thousand and ten thousand people.

Consolidation led to fundamental changes in Hohokam society, to
walled compounds and higher platform mounds—to a more closed
society where the social order had apparently become more exclusive,
more privileged. More than a hundred platform mounds and other
structures now served as temples, elite dwellings, and symbols of power.
There may have been changes in property rights, too, as control over irrigation became more centralized and the leadership consolidated its
power.

As the social order changed, environmental pressures appear to have
intensified. Just how, we don't know, but the massive irrigation systems
no longer produced the food surpluses to support a now much more
elaborate Hohokam society. The collapse came around 1450, probably
a rapid dispersal, household by household, as people moved away to
settle with kin or farmed on a much smaller scale. Their successors
were the Akimel O'odham, probably their descendants, who built their
more modest irrigation works atop those of earlier times. And in 1867,
the Swilling Irrigating and Canal Company used the inspiration of

long-vanished irrigation farmers to found small-scale farming communities that would develop into today's megacity, built on the foundations of one of history's most brilliant irrigation societies. One wonders what Hohokam canal builders, fresh from scraping the desert for months to provide a trickle of water for their fields, would think if they were to fly over modern-day Phoenix, with its sparkling blue swimming pools and sprinklers bathing impossibly green golf course fairways. They would probably be horrified.

CHAPTER 5

The Power of the Waters

THE TERRACES UNFOLD ACROSS THE LANDSCAPE, masking hillsides, following contours, crowding one against the other in an orderly array. New rice shoots poke through translucent, shallow water, soon to grow into tightly packed fields of vivid green. Here and there, a farmer walks carefully along a narrow path between the plots, perhaps balancing with outstretched arms. Terraces, hillsides, gullies, and water channels—the densely cultivated paddies and groves of palm trees extend toward the distant mountains, an irrigation landscape centuries in the making.

High above the rolling fields, the Pura Ulun temple stands serenely on the shore of Crater Lake (Lake Bratan), the realm of the goddess Dewi Danu. The main temple, with its eleven-story *meru* tower (a symbolic world mountain) and the shrine to the goddess, lies amid the water, the place where the "power of the waters" resides. Mountain peaks and towering clouds form the mystic backdrop for the great pagoda; morning mist hovers close above the quiet water. It is here that the irrigation systems of Bali begin, self-sustaining systems that are bewildering yet logical in their complexity.[1] In Bali, the task is not to eke a living from an arid landscape, but to manage a dizzying abundance of water—dozens of inches of rain fall here in the monsoon season.

That memorable and often controversial anthropologist Margaret Mead once wrote of Balinese villagers that their lives were "packed with intricate and formal delight."[2] As part of this delight, they live with and manage water, in a complex melding of practical agriculture and social behavior that epitomizes the complexities of the human relationship with a priceless resource. The same complexity extends to village temples,

Figure 5.1 *The Pura Ulun temple. (Simon Gurney/iStockphoto)*

which lie at the center of Balinese life. Each village has at least three temples: a shrine that revolves around ancestor worship; a death temple, which handles death and the evil forces associated with it; and a water temple, arguably the most important of all—a shrine concerned with the spiritual oversight of irrigated rice fields. With water, as with everything else in Balinese life, ties of ancestry and the forces of the spiritual world are of paramount importance. They come together in an intricate mosaic of rice terraces and irrigation works that have flourished for more than a thousand years.

BALI LIES SOUTH of the equator in the Indonesian archipelago, a few miles east of Java. A curving line of volcanoes forms the backbone of the island, for Bali is part of the Ring of Fire. This makes for a rugged landscape where minor eruptions constantly rearrange the topography with fresh lava flows and ash falls. Most Balinese live within a south-facing natural amphitheater created by these volcanoes that measures no more than about fifty miles (eighty kilometers) across. It is here that irrigation supports thousands of subsistence farmers.

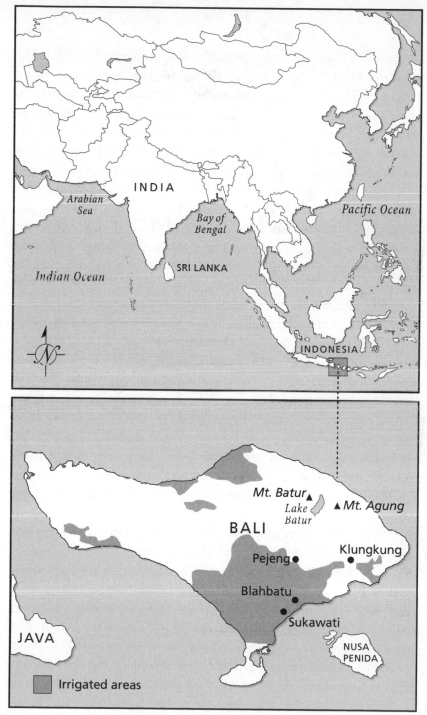

Figure 5.2 *Map of locations in chapter 5.*

When the Balinese look upward toward heaven, they see cloud-mantled Mount Batur, the volcano that dominates their lives. Its crater contains high-altitude Lake Bratan, at 4,038 feet (1,231 meters) far above the elevation where one can grow rice. The lake covers 4,250 acres (1,720 hectares), an immense natural reservoir, whose springs and rivers provide irrigation for all of central Bali. Balinese priests describe the lake as a sacred *mandala* (universe) of water fed by springs lying at each wind direction. The still-active caldera, with its clouds of smoke, represents the *mandala*'s zenith; the depths of the lake, its nadir. Each of the springs around the lake is considered the origin place of waters in different hydrological regions. The priests call the lake a freshwater sea, filled with life-giving water, a sharp contrast with the ocean far below.

"First is the god who reigns in the Ulun Swi temple, who cares for the life in the rice terraces."[3] Balinese water management lies in the hands of the deities. The offerings and rituals in their temples along the rivers that flow from Crater Lake are central to the successful irrigation of thousands of acres of rice terraces. Dewi Danu, the "Goddess of the Lake," "makes the waters flow, those who do not follow her laws may not possess her river terraces."[4] Legends tell us that she and her male counterpart emerged from an erupting volcano and took possession of the land and waters of the island. Dewi Danu rules Mount Batur and the lake, while the god rules the highest mountain, Mount Agung. The goddess has no political associations with Bali's many kingdoms. Her congregation comprises several hundred *subaks*, who make regular pilgrimages to her lakeside temple, the second-most-important religious center on Bali, after that of the god of Mount Agung. (For terminology, see "Some Balinese Irrigation Terminology" box.) Unlike local water temples, Pura Ulun, or the Temple of the Crater Lake, is always open, perennially manned by twenty-four priests, selected in childhood by a virgin priestess. A single high priest, the Jero Gde, leads the priestly hierarchy and is believed to be the earthbound representative of Dewi Danu.

Like his subordinates, the Jero Gde is chosen in infancy after the death of his predecessor. A virgin princess goes into a trance. Possessed by the voice of the goddess, she names the new Sanglingan, or "Lightning-struck," another name for the Jero Gde. The new high priest assumes

Some Balinese Irrigation Terminology

Agama Tirtha	"The religion of holy water."
Bedugl	A small shrine at the point where irrigation water enters a farmer's land.
Subak	A complex of rice fields receiving water from a single irrigation conduit, made up of one or more *tempeks* (see below). In human terms, a *subak* is an association with both secular and religious functions. *Subak* members carry out maintenance on canals and temples.
Tembuku	A wooden water divider that defines a *tenah* (see below).
Tempek	A block of rice fields.
Tenah	The amount of water that will irrigate a set unit of land, usually within a fixed period of time, measured in the *tembuku* (see above) at the upstream end of the field.
Tika	The calendar of cultivation cycles followed by *subaks*, usually kept on painted sticks, a grid of thirty seven-day weeks. The Balinese also have a linear lunar calendar.
Tirtha	Holy water, the common element in all Balinese temple rituals.

his office while still a boy and undergoes purification rituals. Throughout his life, the goddess guides his conduct and decisions. He makes sacrifices to the goddess on behalf of the *subaks* and receives guidance from her in his dreams. The Jero Gde wears his hair long and always dresses in white, the color of purity. His domain is the water temples, sanctified by his relationship to the Goddess of the Lake, which gives him a unique authority over irrigation water. The anthropologist Stephen Lansing, who has made the definitive study of Balinese irrigation, quotes the Jero Gde: "It is only she, the Goddess of the Lake, who can properly give water. She already embodies, incarnates water, which she gives to her *subaks*, from the lake."[5]

The Jero Gde has authority over water allocation. He bridges the living or visible world and the realm of the intangible, residing high above the irrigated landscape and symbolizing the vital relationship. The Jero Gde represents primordial divine power that goes back to, and precedes, creation. As Lansing remarks, "He is not only the First Human. He is also an icon of the temple itself, the social world that continues to originate from the ever-flowing rivers of the lake."[6] Every high priest comes from an ancient mountain lineage known as the Paseks of the Black Wood, which traces its origins back to the time when the supreme deities took possession of Bali.[7]

Balinese water management depends on a close interdependency—a hydro-logic, if you will—between water temples and the *subaks* that visit them, composed of the farmers who are the Goddess of the Lake's congregation. The Jere Gde's authority stems, ultimately, from the complex internal logic of the water temple system. He is, quite simply, the summit of central Bali's water temple system. He symbolizes the "power of the waters."

How, THEN, DOES the irrigation supported by the lake work? The uppermost rice terraces are about twenty-five miles (forty kilometers) from the ocean, so the area of intensive cultivation is remarkably small. Clifford Geertz writes of this densely settled area: "If ever there was a forcing house for the growth of a singular civilization, this snug amphitheater was it; and if what it produced turned out to be a rather special orchid, perhaps we should not be altogether surprised."[8]

This is a land of rivers, about eighty of them in the rice bowl alone, fed by springs and heavy monsoon rainfall. Several hundred inches of rain fall on most of Bali during the monsoon season. Dense tropical forest mantled much of the island until it was cleared for rice terraces and now only survives in the western third of the island, which is unsuitable for irrigation. The torrential monsoon rains and floodwaters have cut deep into the soft volcanic rock, carving channels that flow as much as two hundred feet (sixty-one meters) below the surrounding terrain. It is here, in this dissected terrain, that rice farmers have transformed the landscape for more than ten centuries.

The history of Bali's farmers goes back much further than that, to a period of maritime settlement that brought rice agriculture to the Southeast Asian islands from southern China and Taiwan, perhaps between five thousand and six thousand years ago.[9] The newcomers, often called Austronesians, were farmers and fisherfolk who were expert seafarers. Most of their descendants settled other islands and eventually colonized the offshore Pacific. But those who arrived on Bali turned inward and developed their own distinctive culture, while trading extensively with their neighbors to the east. Their trade routes brought metallurgy and new ideas to the island over many centuries, which were an important factor in the development of the first Balinese kingdoms.

Quite when rice agriculture began on Bali is an open question. The earliest radiocarbon dates for rice husks date to about twenty-six hundred years ago, but the plant may have arrived much earlier. The few radiocarbon dates are for sites occupied perhaps a thousand years before the development of irrigation works. People then grew rice in swamps formed by monsoon rains rather than on paddies fed by irrigation. Rice appears to have arrived before metallurgy, which is not surprising, given that neither copper nor tin occurs on Bali. By the late first millennium C.E., farmers were making common use of metal, especially to forge the digging implements that sliced through the volcanic rock and allowed them to excavate the tunnels needed to carry water from river channels to their fields. We know that such tunnels were in existence by the eighth century C.E., when royal inscriptions refer to such activity, as well as to rice harvests.

Did Balinese irrigation, in fact, begin with tunnel construction, a seemingly laborious way to acquire water for one's fields? Probably not, for the first rice irrigation was, most likely, a logical extension of earlier rainfall-dependent farming. In south-central Bali, a spring named Tirtha Empul produces several hundred gallons of freshwater a second, which flows freely down a valley. By channeling the waters of the spring, farmers could irrigate several hundred acres of carefully terraced landscape in a valley just north of the settlement of Pejeng. It may be no coincidence

that the landscape around the village abounds in archaeological remains, for this region may have been the site of the first kingdoms to
develop on the island.

Pejeng is also home to a gigantic, elongated kettledrum over six feet
(nearly two meters) high, often called the Moon of Pejeng. The drum
recalls great bronze drums from elsewhere in Southeast Asia and may be
as much as two thousand years old. According to Balinese legend, the
drum was the wheel of a chariot that pulled the moon through the sky.
It came loose as it passed over the village and fell to earth, glowing as
brightly as the moon before cooling and becoming a still-revered sacred
object. Whatever its origins, the sacred associations of the Moon of
Pejeng supports the hypothesis that rice irrigation began here first, then
became elaborated as the local population grew, perhaps over more
than a thousand years. The demand for rice, the staple of local diet, was
such that the farmers had to expand their fields and, inevitably, the
waterworks that supported them.

How, then, to enlarge the water supply for farming? Thanks to periodic influxes of nutrients to the fields from incoming river water, Balinese
rice cultivation is unique in that the same field systems have remained in
use for many centuries, without the gradual loss in productivity that
results from decreasing soil fertility and rising soil salinity, such as occurred in Mesopotamia. Here, irrigation is dependent on seasonal river
flows and springs at higher elevations fed by heavy monsoon rains. Virtually all of Bali's rivers flow not at low elevations near the ocean, where
irrigation would be an easy matter, but at higher altitudes, in deep
channels. The only way to reach water is by building earthen and timber
weirs, then diverting river water into tunnels dug laboriously through
the rock. The tunnels emerge at lower elevations, often as much as 0.6
miles (1 kilometer) or more downstream. Then the water flows through
open aqueducts and canals to the summits of terraced hillsides.

After centuries of farming, the irrigation systems are a veritable
maze of tunnels and canals that pass water through blocks of rice terraces. Just accessing the water is a major challenge, for the farmers also
have to work with monsoon rain cycles, during which rivers can carry as

much as ten times more water than during the dry months. Furthermore, the irrigation works have to be flexible enough to cope with everything from flash floods to mere trickles of water at the end of the dry season. They must also be capable of connecting to neighboring systems from different weirs, so that unused water can be passed to another terrace system or returned to the river.

All this requires extraordinarily precise water management, to the point that there are distinct "pulses" of water sent through the system from upstream during the dry months. These pulses carry doses of nutrients, providing a major advantage over irrigation systems where the nutrient supply is constant and unchanging. The result is far higher productivity. Stephen Lansing tells us that the water pulses change the pH value (acidity) of the soil and create cycles of aerobic and anaerobic conditions that circulate mineral nutrients, foster the growth of nitrogen-fixing algae, and determine the activities of microorganisms. Weeds are inhibited; the soil temperature is stabilized; and over the long term, nutrients are leached out of the soil. For instance, both potassium and phosphorus content, essential for rice cultivation, benefit from water pulses.

Balinese farmers mainly grow rice, but their paddies also produce important animal protein such as eels, frogs, and fish. Large populations of ducks thrive on the terraced fields, but they have to be managed carefully, as they adore young rice shoots. After the harvest, the farmers drive them from field to field, where they devour leftover grain and eat some of the insects that could attack the next rice crop. The farmers only harvest the seed-bearing tassels, often leaving the stalks to rot and decompose in the water, thereby adding more nutrients—if they do not cause the field to dry out, which carries the danger of losing nutrients.

The irrigation systems only thrive if all the farmers in the area cooperate throughout the year. They have to coordinate the harvests, make decisions about draining fields and controlling pests, and decide how much acreage will be left fallow and for how long, a critical decision that depends on the kinds of pests found in the fields. Balinese rice farming depends on far more than just synchronizing harvests. It involves careful management of irrigation cycles and water supplies to control pest populations. Much larger social units than a single family or a village are involved when such fundamental issues as pest control

Figure 5.3 *Bali rice terraces. (George Clerk/iStockphoto)*

through carefully coordinated fallowing and water distribution come
into play.

OBVIOUSLY, NO INDIVIDUAL *subak* can set schedules and cropping patterns for every one of them, and it is here that water temples have a role.
Lansing cites the example of the Er Jeruk temple, which lies among
river terraces below the village of Sukawati. This community receives
water from three dams on two rivers that irrigate other terraces upstream before water reaches Sukawati and its nearly 1,000 acres (403
hectares) of terraces. Altogether, thirteen *subaks* receive water, divided
into three groups that are irrigated in rotation. All of Sukawati's farmers are members of the Er Jeruk temple's congregation. All decisions
about planting, water distribution, and fallowing for pest control are
made collectively when the *subaks* assemble at Er Jeruk. Once the dates
are fixed, the decisions are promulgated in the smaller water temples
maintained by each *subak*. Time controls every element—water, planting, and the fate of rodents.

The three groupings of *subaks* plant rice at least once a year during the rainy season. The rotation system takes hold during the dry months. One group is guaranteed a second rice planting, the second can grow vegetables that receive water every five days. The third group plants either the one or the other, depending on whether enough water is available for rice cultivation. In this way, the temple optimizes water sharing, while also ensuring that many fields are left fallow as a way of reducing pests.

Each farmer relies on a basic unit of Balinese irrigation: the *tenah*. This is the amount of water that will pass through a wooden water divider called a *tembuku*, measured at the upstream corner of the field, where water flows into the terrace, usually within a fixed period of time. A *tenah* is not only a unit of shared water but also the amount of water that will irrigate a *tenah* of land. If the farmer plants a *tenah* of seedlings, he will grow a *tenah* of rice.

A series of large dividers lying above each field measures out larger shares of *tenah*. More important, there are diversionary gates and side canals that handle floodwaters from major storms and other unusual events, for excess water is, if anything, a more serious problem than too little. An entire crop can be washed out in hours.

Blocks of terraces form *tempeks*. The farmers within each block share responsibility for maintaining its irrigation works. Effectively, they are secular work groups, with no ritual powers. One or more *tempeks* form a *subak*, which has responsibility not only for its irrigation works but also for the well-being of its rice terraces and their continued productivity. The offerings and rituals in a *subak*'s water temple promote growth and fertility, giving the *subak* important spiritual as well as practical responsibilities.

The Sukawati system is relatively simple compared with some others, among them one administered through the water temples of Kedewatan, halfway up Mount Batur. Here, seven *subaks* share water from a single large canal that flows from a major weir 2.5 miles (4 kilometers) upstream. Ulun Swi, or "Head of the Ricefields," is a major temple located at the point where the waters of the canal enter the terrace system. Just downstream, the canal splits into two. A lesser temple stands by the upstream branch canal, and another one lies about 0.3 miles (0.5 kilometers) downstream, at the point where another branch canal enters a second set of terraces. Each of the *subaks* is a member of one of the two temples; all of them form the congregation of Ulun Swi upstream. At the time Lansing carried out his study, 1,775 congregation members irrigated 1,379 acres (558 hectares) of rice fields. During the rains, the entire congregation planted the same variety of rice to ensure a universal fallow period after the harvest. Here, each *subak* temple chooses what crops to then plant and administers a rotation of irrigation if need be. Each *subak* takes turns undertaking maintenance work on the irrigation works and performing the annual rituals that unfold at the temples. Thus, the daily management of irrigation works is deeply embedded in an intricate social milieu and a hierarchy of water temples.

In Bali, irrigation does far more than supplement the water that comes from rainfall. Here, the farmers use carefully managed water as a way of constructing an artificial pond ecosystem that imposes very severe restrictions on water management. As Lansing points out, the cycles of wet and dry phases govern the fundamental biochemistry of the paddy ecosystem. The 172 *subaks* along the Oos and Petanu rivers use water obtained from rivers and streams flowing through rugged volca-

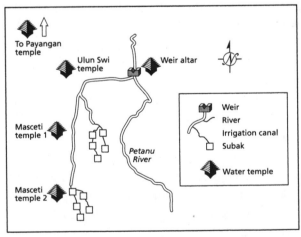

Figure 5.4 *The irrigation system downstream of Ulun Swi temple. (From J. Stephen Lansing,* Priests and Programmers. *Copyright © 1991 by Princeton University Press. Reprinted by permission of Princeton University Press)*

nic terrain. Very few of them can farm successfully with only one weir—and those that do live at the highest elevations. Almost all depend on supplies released by neighbors living upstream for much of their irrigation water. Lansing refers to this as "hydrological dependency," for the survival of the complex water-delivery systems, with their weirs, canals, tunnels, and aqueducts, depends on a system of control that extends far beyond the boundaries of a single *subak*'s rice fields. This control system links watershed and river to weir, weir to tunnel and canal, and village to village in a web of interconnectedness that can be maintained only by intricate social controls. In Bali, as in many other subsistence-farming societies, agriculture is both a social and a technological process. At the very foundation of this process lies water.

THE TRADITIONAL RULERS of Bali were divine kings. Each was what Clifford Geertz once called a "universal monarch, the core and pivot of the universe."[10] There was no supreme leader but dozens of them, all claiming divinity, each with their royal rituals. Balinese kings were all-powerful, and yet they were not, for the rituals of the water temples lay far from the realm of politics and the sedulous recitations of court priests. Water temples lay in the background, inconspicuous, buried in the shady haze of village rituals and folk stories. However, Balinese agriculture and water control were anything but hazy, lying firmly in the domain of the practical realities of subsistence agriculture.

Balinese institutions of all kinds, from households to entire kingdoms, have their own temples and shrines, where offerings are made to the deities concerned with their affairs. For instance, all Balinese markets have temples, dedicated to the goddess Maya Sih, the mistress of illusions, but also to other divinities, like the Rice Goddess Dewi Sri. In a similar way, water temples boast an array of shrines with anthropomorphic deities that define the cosmological and social role of the temple. Lansing points out that the rituals performed at a water temple define its role in local society, in daily agricultural life. The regular ceremonies conducted within its precinct are the way in which the congregation forms task forces to carry out essential tasks, like maintaining weirs, or to decide what crops will be planted and when. In other words, the temples foster voluntary social cooperation between different villages and within society as a whole. Their decisions and rituals are the bedrock foundation of daily life, of short- and long-term survival.

Think of the water temples as links in a chain that connects different parts of the natural landscape with the human groups who seek to control them. Each shrine or temple has an association with some component of the irrigated landscape. The chain begins with a spring or a weir in a river. Both yield water, which is diverted into a tunnel or a canal. This is where the first shrine lies. All the farmers who use this particular water source are the temple's congregation. The presiding deity is the Bhatara Empelan, or the "Deity of the Weir." There is also Dewi Danu, Goddess of the Lake, who, the farmers believe, makes the water flow.

The irrigation canal reaches a complex of terraces, perhaps just over 0.5 miles (1 kilometer) downstream. This is where Ulun Swi, or the "Head of the Rice Temple," lies, sharing the same congregation as the weir temple. The influence of the god Ida Bharata Pura Ulun Swi, the "Deity of the Ulun Swi Temple," extends to all the terraces watered by the canal. There is not much to see—simply a walled courtyard with a shrine, where the farmers make their offerings to the presiding divinity. The offerings at the Ulun Swi temple are an acknowledgment that the farmers depend on the water that flows into their terraces and on the links in the chain further upstream—the waters that come from the weir and the river flow that fills it.

Everywhere, water temples lie upstream of the water system that they control. There are shrines at locations seen to influence every aspect of the terrace ecosystems, from weirs to fields. They provide the link between the people and their *subaks*, for the gods worshipped in the shrines control the irrigation systems and their water supplies. For instance, the weirs are humanly constructed, but become parts of the physical landscape. The weir god is an anthropomorphic being, whose image draws attention to the critical link between humans and the artificial landscapes that they have created. In a real sense, the weir, the realm of the gods, is a social unit presided over by the weir god, a permanent institution that transcends the transitory lives of the individual farmers who obtain water from it.

There's a second pervasive relationship, too, for the hierarchy of temples is interdependent along the length of the irrigation system. Thus, when major annual festivals take place at the Head of the Rice Temple downstream, the Deity of the Weir and the gods of other upstream temples are invited downstream to receive offerings. Every temple has shrines that commemorate symbolically the gods from other temples, so offerings at each one can celebrate the interdependence between them.

The hierarchy of shrines descends to individual fields. Every farmer maintains a small shrine, a *bedugul*, at the point where irrigation water enters his land. This is the sacred part of his holding, the place where he makes offerings to the Rice God, symbolized by his crop.

Upstream and downstream temples have different symbolic functions. Upstream water nourishes the land and is considered a gift from the Goddess of the Lake. Downstream water cleans and purifies, and washes away pollution. Upstream, temple priests collect water in sacred vessels—treating it as holy water, as befits a gift from a deity. The downstream flow remains in the system and washes away ashes from sacrifices and other impurities, such as sewage, which the rivers transport to the ocean. There is enough flow to allow the collection of pure drinking water from springs upstream. Some human waste undoubtedly ends up on irrigation fields, where it serves as fertilizer, while platoons of Muscovy ducks waddle across the thinly flooded fields before planting, eating anything they come across. There is powerful symbolism surrounding the waters. Those upstream are the life givers, while those at lower elevations have important associations with dissolution and regeneration, whereby the human detritus carried down by the rivers enters the sea and is dissolved, returning to the wild, natural state.

THE GODS IN the water are agents of purification. They control the routes by which they nourish and then purify the irrigated landscape delineated by water temples that extend right downstream to the borders of the ocean.[11] However, the relationship is more complicated than merely that of unchanging rivers and fixed irrigation systems. The Balinese call their religion Agama Tirtha, or the "religion of holy water." *Tirtha*, holy water, is common to all Balinese rituals, whether performed in water temples or not. Every act of worship culminates in a blessing of holy water, which is sprinkled on one's head. The worshipper also drinks a few drops of it. Libations of *tirtha* are poured over offerings, sacrifices, buildings, and irrigated fields. As Lansing points out, holy water links upstream and downstream, for it is both a blessing and a purification.

Upstream is preferable to downstream: Every irrigation farmer knows the advantages enjoyed by those whose fields lie upriver or up-canal. *Tirtha* acquires its holiness from its position on the river. Every Balinese water temple has its own holy water, endowed with unique qualities that define the temple, its deity, and the congregation. And this water comes from upstream, eventually from the weir where wild water enters the irrigation system and becomes subject to human control. One can imagine a priest from the weir temple collecting cupfuls of *tirtha* from the waters above the weir. He places them before the weir shrine, making offerings to the god to sanctify the holy water. Once sanctification is achieved, the priest stores the cups carefully, for the *tirtha* retains the sacred qualities of the waters immediately upstream. At the Ulu Swi temple on the Petanu River, farmers from *subaks* downstream come to ask the priests for holy water from the Bayad weir. They make offerings, receive the water from the altar near the weir, then carry it back to their own temple. There, they mingle the newly acquired holy water with water from their own temple and sprinkle it on their offerings. Each temple has its own unique holy water. But water from the Bayad weir is of no interest to farmers cultivating along a different channel of the river.

The holiest water comes from far upstream, water so sacred that it is treated like a god and may even represent the divinity that created it at offerings outside the temple. *Tirtha* creates a complex web of relationships between social groups that becomes ever larger as one travels upstream. Holy water "flows" down the irrigation systems and defines the origins of each such group, as well as the relationships between water temples. At the pinnacle, as it were, lies the Temple of the Crater Lake, on the volcano's rim, where the entire process begins.

The Tenth Month, March—when the rainy season ends. It's time for the preparation of *tirtha* for the farmers' fields, for the "pregnancy of the rice." Clouds mantle the volcano's cone as the gray dawn gives way to the rising sun. A solemn procession of Crater Lake priests wends its way along the steep path up the volcanic cone toward the summit. Brightly colored figures appear and disappear in the drifting gray, the clouds merging with the steam issuing from the vents at the summit. The delegation prays and makes offerings to "request holy water." Meanwhile, the senior priest collects water droplets that have condensed on the nearby rocks from the steam. With reverent care, the functionaries carry the filled cups of the most sacred holy water down to the Temple of the Crater Lake, where they mingle it with holy water from the eleven streams around the lake.

Later in the Tenth Month, large crowds gather. Delegations from over 200 *subaks* arrive, each with its own sacred vessel. They offer prayers and make offerings; the temple priests pour about a quart (one liter) of Crater Lake *tirtha* into each vessel. Each group of delegates carries its share home with great care, then mingles it with the waters of its own regional water temples. Finally, the local priests distribute small samples to individual farmers, who sprinkle them at the upstream edge of their fields. In this way, each farmer and his fields enjoy a close symbolic link both to upstream temples and to the ultimate water source. *Tirtha* has strong associations with upstream—the place of origin. In the hands of priests and *subaks*, it flows downstream to the fields and villages.

Holy water provides a set of symbolic connections, as do offerings to the gods, those of the villages associated with all the temples, ancestral deities, and the Ratu Ngurah, Lord Protector of the Earth. These wider connections are important, for they define the interdependent relationship between irrigation and wider society.

Every decision about planting and other agricultural concerns begins in the water temples, where the *subaks* that form their congregations meet. As the village head of Sukawati told Lansing, "every new planting season there is a meeting. If the planting schedule is not to be changed, there is no meeting. Of course, the ceremonies held here go on regardless, a one-day festival every six months, and a three-day festival every year . . . This place is the home of the spirits of those who have preceded us, who built this temple."[12]

A Balinese farmer's year unfolds in a sequence of field rituals that mark such events as "water-opening," transplanting, flowering, and harvest, but the productive cycle is more complex than this. About every ten years, the Ulun Swi temple organizes an "opening of the waters" ceremony. On the appointed day, representatives from each *subak* in the congregation bring offerings to the temple. One can imagine the clouds of incense and the chanted prayers as the temple priests invoke Dewi Danu, the Goddess of the Lake, the deity of the Temple of the Crater Lake, and the "Deity of the Masceti Temple of Payangan." Payangan lies upstream of Ulun Swi, a focal point for another cluster of *subaks*, whose cropping sequence affects water flow to Ulun Swi. As the prayers continue, a small group of *subak* representatives walks along the banks of the main irrigation canal to the altar by the river weir. There, they lay out more offerings, both for the visiting divinities and for the god of the weir. Excitement mounts as holy water from the Temple of the Crater Lake is poured into the entrance gate of the main irrigation canal. Thus, every ten years or so, the *subaks* use holy water to acknowledge their reliance on the weir, their Payangan neighbors upstream, and the Temple of the Crater Lake.

Everything fits into a series of cycles, defined by a cyclical calendar known as a *tika*, what Lansing calls "a wheel within wheels view of time." The farmers prepare their fields, then transplant the rice seedlings into the fields, each of the *subaks* setting a seven-day period for its members to transplant, a process that has to be coordinated with the flooding of the terraces. Each farmer has his own cycle of planting and harvest, which interlocks with the *subak*'s collective cycle. The aggregates of these cycles mesh with those of larger temples. The *tika* is a wooden or painted calendar that defines a grid of thirty seven-day weeks, with other elaborations that allow farmers to synchronize the timing of water use and planting. All the farmers know its sequence of days and weeks by heart.[13]

THE BALINESE LIVE in an engineered landscape that has taken shape over many generations, a world where society and nature have grown interdependent. Each generation has managed the engineered landscape, and the process of management has shaped social relationships for each one. The advent of Dutch colonial rule, and especially of the Green Revolution, which came to Bali in 1971, brought in new practices, including the widespread use of chemical fertilizers, major shifts in cropping practices, and, at one point, a prohibition on traditional rice varieties. Many of these innovations coincided with a push for self-sufficiency, and they led to abandonment of the *tika* cropping calendar. Continuous cropping effectively removed control of irrigation and farming activities from the temples. The inevitable result was chaos in water scheduling, an explosion of pests, and a constant race between farmers and pests centering on the latest pest-resistant rice forms. Now a centralized bureaucracy controlled, or rather, tried to control, irrigation agriculture, and it simply did not work. The traditional practices are still in use and work well.

Balinese farmers look at their environment and their farming through the prism of the interlocking cycles of the *tika*. Mesopotamian rulers centralized control of their farming, and Chinese emperors unleashed thousands of workers on flood-control projects, but in the case of Bali, such centripetal administration could not have worked. The Balinese prefer a complex orchestra of meshing cycles that worked with brilliant success for many centuries. It was no coincidence that Balinese rice farmers achieved self-sustainability. They believed that the deities were on their side, and they still do today.

Waters from Afar

Labor, power, and early water management: In which we explore the beginnings of larger-scale irrigation that gathered water brought by rivers from afar. Part II describes the village farmers and early cities of Mesopotamia, the flood irrigation of the Egyptians, and the increasing complexity of Assyrian and later water management. The Greeks and the Romans built on the expertise of earlier civilizations with aqueducts and urban water systems. In the end, water begins to become a commodity, even if the gods provided it.

CHAPTER 6

Landscapes of Enlil

NORTHERN IRAQ, 7000 B.C.E.: Gray clouds, heavy with rain, hang low over the mountains. Their ragged edges swirl across forested ridges as shower after heavy shower advances across the jagged peaks. Fierce thundershowers have buffeted the high ground for days, but only a few sprinkles have fallen on the plain. Down in the village, the people listen to the runoff flowing down the nearby wadi. At first, the sound is soft, even soothing, but the noise level soon escalates. Pebbles, stones, and then boulders roll downstream; tree trunks from far uphill roll down the ravines like giant spillikins. The din of the flash flood is like music to the farmers' ears, a portent of water for their fields. They've prepared for this moment for months.

As the floodwaters rise, the elders move upstream to the place where the village has tended stone-and-brush check dams since long before they were born. A barrier of water-worn boulders extends across the wadi, capped with dense brush piled thickly on the upper side. Here, the water is deeper and pools for some distance upstream. Two shallow channels divert floodwater from the deepening stream as the rising torrent turns the dam into a weir. A group of young men stand waist deep in the water, casting boulders onto a weak spot on the dam. Without warning, a rapid surge of floodwater sweeps downstream. Large boulders rumble menacingly along the riverbed. The youths scramble for dry land. One of them falls in deep water as a jagged rock knocks him senseless. Before anyone can move a finger, a breaking wave sweeps the unconscious young man over the submerged dam. His limp body vanishes with the inundation as his companions chase along the bank, to no avail. Hours later, the badly smashed corpse grounds far down the wadi, almost unrecognizable.

The older men follow vainly and shake their heads. They have known years when the boulders gave way and their carefully nurtured pool vanished in short order. They also remember other storms when young men tried to stem the flow and were washed downstream, knocked unconscious by rolling boulders. Every year, they warn a new generation, sometimes to no avail. This time, the rocks hold, as they have most years after generations of siege from upstream. Without a further word being said, the elders follow the edge of the rising water and watch their sons as they keep water flowing into the channels leading to the fields.

The farm plots lie slightly downslope. They are square or rectangular, each carefully laid out with a minimal gradient, enclosed with low earthen embankments piled highest at the lower end. The floodwaters pass smoothly down the diversion channels, then onto the waiting fields, rising slowly until each is completely inundated. Young men and women stand guard over the narrow breaks that divert water onto the carefully prepared land. The older men watch the process carefully to ensure that everyone receives as much water as possible. They keep a close eye on the rain-shrouded mountaintops and an expert ear cocked for the decibel level of the flood, for they know that the water can recede as fast as it rises. This time, they are lucky. The waters rise far above the dam and stay there for several hours. By the time the boulders appear above the surface and the runoff eases, the closely packed grid of square fields is completely under water.

Not that the work is over. The villagers squeeze all the advantage they can from the flood. They top up the fields as often as they can for as long as the diversion channels flow. By the time the wadi resumes its normal trickle, the soil in their plots is muddy and waterlogged. The ground steams in the sun as the waiting begins anew. The farmers have waited for water. Now they will begin planting just as soon as the soil begins to dry up and seed can germinate in the damp earth . . .

Pity farmers who dwell in arid and semiarid lands, where rainfall is seasonal, and usually unreliable, and permanent water supplies are a rarity. Between 15 and 20 percent of humanity dwells in such environments, where our predecessors developed uncanny skills at conjuring water

from seemingly waterless landscapes. Their expertise is even more remarkable when you reflect that they could acquire water only from two
sources. They could find it within the arid landscape itself, by tapping
groundwater through simple wells or water holes. However, more reliable supplies usually came from outside. People living near mountain
ranges and steep hillsides depended on runoff from higher ground. Then
there were rivers large and small, among them the floods of the Indus,
the Nile, the Euphrates, and the Tigris, where water that inundated natural basins could be diverted into natural or artificial channels to irrigate
nearby fields.

The arid and semiarid Near East is the cradle of cereal cultivation
and animal husbandry, with a history of successful farming in dry, in-
tractable environments that goes back to as early as 10,500 B.C.E. Farmers
dwelling here lived in a world of many months when the land was dry,
then a few precious ones when the landscape came to life and featureless

Figure 6.1 *Map showing locations in chapters 6 to 8.*

plains turned green with native scrub and wildflowers, thanks to sporadic rainfall. The growing season was brief and magical, the work, people believed, of powerful supernatural forces. The Sumerian *Dispute Between Summer and Winter*, written in the third millennium B.C.E., dramatizes the moment: "Summer founded houses and farmsteads, he made the cattle-pens and sheepfolds wide. He multiplied the stacks of sheaves in all the arable tracts. At their edges he made . . . flax . . . ripen."[1]

Even the earliest farmers linked their fields and their harvests to the benevolent and malevolent forces of the cosmos, interceding with the supernatural through the intermediary of revered ancestors, whose figurines and skulls lie in the heart of their villages. But while they might invoke the power of ancestral guardians, there was little they did not know about the properties of their soils and about obtaining water to nourish them. They used natural seeps and springs and were well aware that water flowed downslope and could be diverted onto nearby fields. Like farming itself, simple gradient irrigation on a small scale was not a revolutionary invention. The major changes came when village farmers began to depend on irrigation water and virtually nothing else.

WHEN AND WHERE did such village irrigation farming begin? The answer lies in the natural landscapes of the Iranian plateau and northern Iraq. In environments where standing water is a precious rarity, you can get water by dipping your hands or a container into a river or lake. Or you can tap groundwater lying near the surface by digging water holes and shallow wells. Water holes are the oldest way of acquiring water and were in use by hunter-gatherers thousands of years before farming began. Australian Aborigines and other hunters living in dry environments knew well that water could be obtained by digging into dry streambeds or at often inconspicuous locations visited for many generations.

Most farmers' water holes were little more than irregular pits excavated by hand or with digging sticks, extending a few feet below the surface. As opposed to wells, which were carefully dug and usually lined with bricks, stones, or even wickerwork, such small pits were informal ways of tapping groundwater, used by village farmers long before irrigation came into widespread use. Water holes worked well in areas where groundwater was relatively accessible, but required laborious watering of fields by hand, which immediately limited the acreage that could be cultivated.

Apart from water holes, the primary source of water for many subsistence farmers in the northern Near East was groundwater, provided it was close enough to the surface to be accessible in the absence of pumps and other lifting devices.[2] The commonest sources of groundwater are alluvial aquifers, which, for the most part, lie relatively close to the surface. They occur along major river valleys at the edges of higher ground. Stream flow percolates along floodplains and across the top of alluvial fans, where coarse gravels are to be found. (An alluvial fan is a fanlike deposit formed when a fast-flowing stream reaches a flatter gradient. The water slows, then spreads out, depositing gravel and sand, usually at the base of a canyon as it debouches onto flatter terrain. Such fans often occur in desert and semiarid areas that are subject to flash flooding from thunderstorms on nearby higher ground. Water percolates through the fan, then flows downstream when it reaches a more impermeable layer, emerging at the base of the fan as a spring or a seep.) Most of the groundwater recharge takes place between February and May, when mountain ranges lose their snowpacks, which flow down rivers in well-defined, if irregular, pulses that produce higher water tables at the tops of alluvial fans. Actual rainfall at lower elevations is not a major factor in the buildup, for much of it is lost to evaporation.

Every farmer must have known about alluvial fans, for these are places where water seepages are to be found. For thousands of years, nearby villagers would have tunneled back from the base into the loose gravel of the fan to concentrate the flow in one place, in a pool. In time, the diggers would have started tunneling to follow the water back to its source. Digging such tunnels was hazardous at best. Within seconds of excavators' hitting an exceptionally waterlogged gravel layer, the roof would collapse, burying them without warning.

One can imagine two boys tunneling into sand and loose gravel on their hands and knees, digging into the slope with their bare hands and short digging sticks, kicking the soil back behind them. Water percolates alongside their sweating bodies, as their father offers encouragement from outside. Suddenly, the ceiling collapses in a torrent of gravel and groundwater that buries the diggers in an instant. The father darts forward, grabs each boy by the ankles, and drags them clear. The elder one wriggles frantically and sticks in the narrow entrance. He begins to choke, but is pulled clear just in time. Next time, they'll try digging in another place . . .

Tunnel digging must have been, at best, a chancy way of locating groundwater. Many centuries were to pass before the diggers tried another tactic: sinking a vertical shaft to locate the aquifer before tunneling horizontally downslope through the fan. If the gradient was correct, the village lying downslope of the tunnel would have a comparatively reliable water supply that might fluctuate through the year but was not dependent on annual rainfall. No one knows when such well and tunnel digging began, but as we shall see in chapter 8, these simple constructions developed into the *qanat*, one of the simplest and most enduring inventions of history.

If you cannot locate water close to where you live, then you have to bring it to you by diverting it from lakes, rivers, and streams. Almost invariably, farmers in the northern Near East using groundwater also relied on irregular rainfall, limited amounts of water diverted from streams, or seasonal runoff from higher ground. They could store surplus rainfall or excess runoff in underground cisterns, dammed riverbeds, or natural flood basins. The storage pools behind them were normally tapped from the top. Silting and shallowing became a problem, and they were soon abandoned. Numerous abandoned dams littered most arid landscapes where they were used. Many communities also relied on the simplest form of irrigation farming: inundation agriculture.

Inundation farming is exactly what the term implies: flooding fields and letting the water and its nutrients soak into the soil, then planting your crops in the damp earth as it dries. This simple and effective technique was a staple of ancient Egyptian agriculture and dates back to the earliest days of farming. The discoveries at Wadi Faynan, in Jordan, described in chapter 2, tell us that many small communities made use of runoff from short-lived storms and flash floods to inundate their fields, which enabled them to sustain small-scale farming for many generations. Inundation worked well in environments where seasonal rainfall was relatively predictable in most years.

Strictly speaking, irrigation agriculture is something distinct from relying on receding, diverted floodwaters, for it involves much more work: storing, distributing, and then draining water for each field system, not merely allowing water to soak into the ground. However, the two form a continuum, with the one probably developing from the other in or around Mesopotamia, "the land between the rivers," between the Euphrates and the Tigris, in what is now Iraq.

THE EUPHRATES AND the Tigris were defining rivers in the history of early civilization. A century ago, many scholars called the territory between them the eastern horn of the Fertile Crescent, a great band of flat and mountainous terrain that arced between the Nile and southern Mesopotamia, the cradle of the earliest civilizations. (Southern Mesopotamia refers to present-day southern Iraq between the two rivers.) Few people talk about the Fertile Crescent anymore, but there's no denying that mountains, plains, and rivers helped define the societies that arose there. Rugged mountain ranges formed barriers and fostered cultural and ethnic diversity. Plains and rivers made communication easier and channeled it in clear directions, nurturing early civilization.

When the Assyrian king Sargon II visited the mountains around Cudi Dag, traditionally the resting place of the biblical ark, in the eighth century B.C.E., "he leapt from rock to rock like an ibex, and then sat on a rock and had a cold drink."[3] The mountains, then mantled with thick forests and teeming with wild beasts, filled Sargon and other lowland monarchs with awe. Their domains lay in wide-open spaces, on plains that extended from the Zagros range to the Persian Gulf, dissected by the great rivers and their tributaries. The northern plains on either side of the Tigris form rolling, often broken terrain with occasional patches of alluvial soil and wadis. The topography militated against any form of irrigation, so the farmers depended on rainfall and shallow groundwater. By 6000 B.C.E., agricultural communities flourished over a wide area, delimited to the north and east by the foothills of the Zagros and Taurus mountains.

The southern and eastern limits of dry farming varied considerably from year to year. Barley, the major staple, required about 9 inches (20 centimeters) of annual rainfall, wheat about 10 inches (25 centimeters), fruit trees considerably more. In practice, about 12 inches (30 centimeters) of dependable rainfall was needed, provided that adequate amounts fell in three years out of every five. So the southern limit of dry agriculture was an irregular line that shifted with even minor changes in rainfall. South of this tattered frontier, irrigation agriculture reigned, even when more rain fell upstream, as was likely in the past.

The lands to the south of modern-day Baghdad form an austere floodplain landscape, almost devoid of rainfall. Extreme forces of nature confront you here on every side—some of the hottest summer temperatures on earth, bitter winter winds and tumultuous storms, sudden river floods that can sweep away a village in moments. Here, nature and the gods were often malevolent, and in later centuries the region's rulers were given to despotism. The Sumerians, who developed the first urban civilization in the south, worshipped ancient gods, among them Enlil, who was "father of the gods" and presided over this realm. It was he who made the day dawn, and he who established plenty in the land: "Without the great mountain Enlil, no city would be built, no settlements would be founded . . . no carp-filled waters would flood the rivers at their peak . . . in the sky the thick clouds would not open their mouths; on the fields, dappled grain would not fill the arable lands."[4]

The floodplain between the Euphrates and the Tigris seems flat, but is never entirely so. Under natural conditions, the rivers would meander over the landscape and sometimes change course when they burst their banks during major floods. The Euphrates has shifted its lower reaches time and time again without warning, transforming the natural and humanly altered landscape in fundamental and sometimes catastrophic ways. Dense vegetation once grew close to the rivers, but as the fresh groundwater of the rivers tailed off, the trees gave way to desert and sand dunes, landscapes where the remains of old river channels, abandoned canals, cities, and villages can be seen, shrouded in windblown sand. As the Victorian excavator Austen Henry Layard wrote of Babylon in 1853, "the plains between Khan-i-Zad and the Euphrates are covered with a perfect network of ancient canals and watercourses, but 'a drought is upon the waters of Babylon, and they were dried.' Their lofty embankments, stretching on every side in long lines until they are lost in the hazy distance, or magnified by the mirage into mountains, still defy the hand of time, and seem rather the work of nature than of man."[5]

An unlikely environment for farming, one might think, but appearances are misleading. Simple gravity-based irrigation provided river water to patches of arid but fertile soils and small farming villages many centuries before the first cities appeared by the Euphrates, during the fourth millennium B.C.E. The landscapes of Enlil had deep roots in the past.

There was never any question of a fully irrigated floodplain, where closely packed, regularly watered fields stretched to the far horizon.[6] The flat topography meant that any large-scale irrigation works would have required a massive human investment far beyond the resources of a Sumerian ruler or his immediate successors. The gradient of the floodplain changes by a mere 98 feet (30 meters) over 435 miles (700 kilometers), which makes any form of gravity-flow irrigation canal—even just the process of enlarging a natural channel—a challenging undertaking. Estimates of the amount of arable land at the time are at best approximations. Only 19,700 square miles (51,000 square kilometers) of land were either cultivated or cultivable. Yet there was only sufficient irrigation water for about 1,600 square miles (4,143 square kilometers) by the most optimistic estimates. The actual figure may have been closer to 3,100 to 3,800 square miles (8,000 to 10,000 square kilometers) of arable land. Large tracts of the river floodplain were uncultivated and formed a mosaic of marshes and lakes, desert, sand dunes, and flood basins, as well as date palm groves. Where the water could not escape, it collected and formed marshes, especially in the extreme south. The wetlands teemed with fish, turtles, and waterfowl; extensive tracts of reeds made invaluable building material.

To farm in the south required not only hydrological skill but also constant vigilance against the ravages of capricious rivers. Both the Euphrates and the Tigris flowed down very low gradients, which resulted in progressively less water, in part because of irrigation upstream and also because of evaporation. Farmers living in the central part of the plain focused their efforts on the slower-flowing Euphrates, which was easier to control. The river entered the floodplain at a higher gradient

than the Tigris and also carried more nutrients than those deposited by
the latter. As soon as it entered the plain upstream of Baghdad, the Euphrates split into at least two branches that also provided opportunities
for diverting water downstream.

The Euphrates deposited its silt load and formed well-defined levees
on either bank. These provided raised strips of coarser, more permeable
soil close to the river, and the slopes provided slightly better drainage
that helped prevent rising salinity in standing water. Judging from today's practice, even the earliest farmers would have planted vegetables
and fruit trees along the levees, using land away from them for more tolerant cereal crops. When the river rose, the water spilled over the levees,
sometimes causing it to change course to a lower level, a process known
as avulsion. Some of these avulsion events were opportunities; others
were potential catastrophes. When megafloods raged downstream, such
as those that occurred along the Euphrates during the second millennium B.C.E., the banks burst on a large scale, clogging major channels
with silt and vegetation and causing havoc, especially at points upstream on the floodplain, where the flow was strongest. In extreme flood
years, the farmers left weirs and flood regulators open to carry as much
water away as possible. *The Epic of Gilgamesh* describes how the storm
god Adad caused a great tempest with roiling floods. But all was well:
"Ninurta [the god known as Lord of the Earth] lets the water flow
through the regulator."[7]

As a result of numerous avulsions, some caused by floods, others by
human action, elongated channels flowed from north to south along the
Mesopotamian plains, many of them simply branches of the main rivers,
others excavated channels, others a combination of both. If they were
natural, they would only need cleaning at strategic points where sediment buildup was unusually fast, or at key bifurcations. Much early
Mesopotamian irrigation resulted from localized, ad hoc diversion of
river water using natural channels. Many centuries passed before any
rulers invested in irrigation works on a large scale. They couldn't afford
it, for they had neither the people nor the food surpluses to issue as
rations.

IRRIGATION IN SOUTHERN Mesopotamia relied on water from far away carried downstream from upland catchment areas by rivers, the largest of them being the Euphrates and the Tigris. The sources were perennial, though the flow varied through the year. The lowest water levels came in autumn, when farmers were carrying out the initial watering of their fields. Exactly where an irrigation system began depended on the gradient, which had to be just right. If the bed was too steep, the water would flow too fast and erode it; too shallow, and silt would accumulate and rapidly fill the channel. Controlling the gradient meant following even minor contours, so that the water flowed across a slope at the correct angle, sometimes at a shallower, and slightly higher, altitude than the nearby river. Even the most basic of irrigation systems were useless unless they were sustainable and easily maintained.[8]

For the most part, early irrigation works used natural channels that took the path of least resistance through the alluvium. They carried both silt and water and soon became clogged with reeds and sediment, as did side channels dug into their banks to lead water to nearby fields. For this reason, many communities tapped rivers during the low-water months, when the reduced flow carried less silt and debris. How much work was involved in maintaining such waterways is unknown. Nor do we know whether neighboring villages drawing water from the same channels would collaborate with one another when the time came.

Even routine maintenance must have been a brutal task in intense summer heat. We can imagine small groups of men wearing little more than loincloths wading knee-deep in soft mud. They work silently and steadily, using both hands to rip out tall reeds from the bottom. When the vegetation is gone, they use their hands to fill baskets with nearly liquid mud and heft them laboriously onto steep piles along either bank. One young man pulls on some reeds too hard, and the roots give way without warning. He topples backward into the muddy water. Everyone pauses for a moment and enjoys the sight of the victim sprawling in the mud . . .

Then there was another difficulty: salt buildup. The salt came downstream from distant mountains and was usually washed away through the permeable soils. Away from main watercourses, soils are less porous, especially when irrigated, so the water table rises and capillary action brings salt to the surface. Ancient farmers combated salinization by planting salt-resistant barley, and also by fallowing. Irrigated fields support a water table about twenty inches (fifty centimeters) below the surface. Nutrient-rich wild plants grow thickly on harvested fields and prevent wind erosion. However, they also draw moisture from the soil, which dries out the subsoil, a process that continues if the field is not planted once again, preventing the water table and salt from rising to the surface. Once the field is irrigated anew, the water leaches salt from the surface and carries it underground, where it is harmless. This fallow system works well for a long time, until the subsoil trap fills and rises toward the surface. Then the ground has to be abandoned for up to a century. However, as the farmers knew, to shorten the fallow period is to court disaster, as happened in later times (figure 6.2).

There was far more to irrigation systems than merely diverting water and allowing it to flow into artificial or natural channels. Controlling the

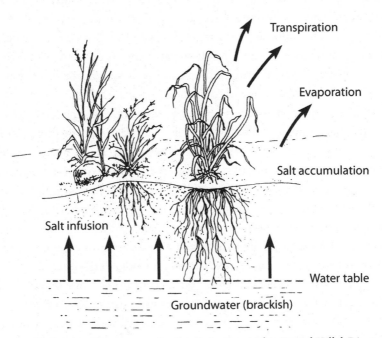

Figure 6.2 *The process of soil salinization. (After Daniel Hillel,* Rivers of Eden, *fig. 3.5. Courtesy: Daniel Hillel and Oxford University Press)*

outflow down the irrigation system went hand in hand with maintaining an adequate supply upslope. Many systems made use of storage basins, blocked by brush or reed weirs, which could be readily breached when water was needed, then easily rebuilt. Such temporary barriers would not, of course, hold back torrential floodwaters, but they appear to have been effective when river levels were lower, during the major watering seasons of autumn and winter. As irrigation became more formalized in later centuries, people turned to baked-brick and bitumen regulators.

The branching and often sinuous irrigation channels passed into the fields (figure 7.1).[9] Fourth millennium survey texts have survived, which show that fields were rectangular, square, or delineated in long strips, but their boundaries, presumably low silt banks, have long since vanished. By that time, the custom of bounding fields was already long established, probably using very simple geometry. The watering itself presented different challenges than those faced, for example, by the Egyptians, who relied on the retreating waters of the summer flood. Their Mesopotamian contemporaries had to inundate their fields in the fall, when the rivers were at their lowest. Thus, efficient use of every drop of water was a necessity, with careful attention paid to fair allocations for all. There must have been times when there was not enough water to go around. Fortunately, cereal crops need only four to six thorough waterings during the growing season, so some system of equitable rotation would have provided sufficient supplies for all. This made the task of coordinating inherited water rights and the fair rotation of precious supplies especially important.

Many Mesopotamian communities lived with the constant threat of water shortages, but they also had to accommodate excessive flow for their irrigation needs. Usually, the overflow passed beyond the field system, often ending up in nearby marshes or rejoining the main river downstream. When it did not, the landowners had to drain off surplus water from their fields to prevent the rising of the water table and salinization.

The routines of irrigation changed little from year to year, century to century. Cities rose and fell, were conquered and rebuilt; rulers boasted of their invincibility, then vanished from history; armies marched and countermarched. But here, as elsewhere in the world, behind this tapestry of stirring but transitory deeds lay the world of the village farmers, which changed but imperceptibly over the generations. They shaped the landscapes of Enlil long before teeming cities rose along the Euphrates.

STUDYING ANCIENT IRRIGATION requires a constellation of skills and strong legs, for you are studying not just single sites but entire artificial landscapes—when you can find them. Only shreds and patches of very early irrigation survive, thanks to thousands of years of flooding and silt accumulation, to say nothing of intensive farming activity. The clues come in many forms: Aerial photographs and satellite images, as well as GPS readings taken from sites on the ground, reveal subtle concentrations of inconspicuous ancient farming villages along riverbanks. Unusually green vegetation may mark the twists and curves of long-abandoned river channels and canals. None of this early irrigation was on a grand scale. It was confined to the surroundings of small villages and was often transitory, as floods shifted and alluvial fans developed new seeps.

It's probably no coincidence that the known early irrigation works clustered around such fans, where groundwater could be tapped. During the sixth millennium B.C.E., the village of Choga Mami, east of the Tigris in eastern Iraq, lay on an alluvial fan within a small oasis near the Gangir River, a place where the rainfall is about eight inches (20 centimeters) a year.[10] The first settlements lay up- and downslope along natural watercourses created by the fan. At first, the farmers merely channeled the water flowing down the main slope. As the population rose, the inhabitants started manipulating both seasonal floods and the river flow along the fan. Later, before 5000 B.C.E., they dug irrigation canals that followed natural contours over wide areas of the surrounding plain, which enabled them to water fields some distance from the river, even if increasingly brackish water made cultivation difficult.

In the Daulatabad region of arid southeast Iran, the local farmers could not rely on rainfall but had to make use of water flow and runoff from ephemeral wadis. Fortunately for science, the ancient landscape survived almost intact until the 1980s owing to a lack of later settlement. Archaeologist Martha Prickett surveyed low mounds, fields, and water channels that developed on gravel fans deposited by the Rud-i

Gushk River above a dry lake depression.[11] The small irrigated fields were up to two tenths of an acre (six hundredths of a hectare) in area, bounded by low stone ridges. In the northern fan area, where the best soils lay, the irrigated land covered some 740 acres (300 hectares), in a place where the soils would retain moisture.

The farmers prospered. By the mid-fifth millennium B.C.E., fields had appeared in the middle fan area, where water came from both local floods and the inundating river. Prickett recorded two possible canals that may have been artificial but more likely were natural drainage channels eroded into the gravel fan, perhaps straightened by the farmers. Irrigation canals proper were shallow depressions up to thirty-three feet (ten meters) in length. They probably received their water from temporary stone dams that deflected water to fields further downslope. This simple but effective irrigation economy produced crops of barley, millet, and wheat for many centuries, until the fourth millennium B.C.E., when a shift in the pattern of monsoon rains, described in chapter 8, caused what may have been a more verdant landscape to dry up.

The Daulatabad landscape shows us how the earliest irrigation may have worked, the farmers using alluvial fans and a combination of simple irrigation and seasonal floods to water their crops. Even the most developed systems were of modest size, the channels rarely longer than about 3.8 miles (6 kilometers). Each community relied on the simplest of methods, channeling water down existing slopes or natural gradients produced by alluvial fans. In the case of Choga Mami and Daulatabad, alluvial fans formed the earliest irrigated landscapes for small villages that probably numbered no more than fifty to one hundred people, linked by ties of kin.

CHOGA MAMI AND Daulatabad flourished during times of higher rainfall, which were soon to end. During the late Ice Age, before fifteen thousand years ago, the Persian Gulf was still dry land; global sea levels were some three hundred feet (ninety meters) below modern levels.[12] The Tigris and the Euphrates flowed through deep valleys into the Gulf of Oman, about 500 miles (800 kilometers) south of their present estuaries. As sea levels rose during the subsequent warming, the newly formed Persian

Gulf caused massive silt buildup in the Mesopotamian plain. Deep sea cores from the Indian Ocean tell us that summer temperatures were higher and rainfall was greater than in earlier millennia between 10,000 and 4,000 B.C.E. Precipitation may have been between 25 and 35 percent higher than today, much of it falling during summer monsoon storms.

More-predictable rainfall, especially in the spring, was a huge boon to farmers the length of the great rivers. In the north, farmers could rely on better crops from fields nourished by more-predictable winter and spring rains. The longer rainy season had even-more-significant benefits in the south. Here, the rains lasted well into early summer, providing an extended growing season helped by the timing of the summer inundation. Today, the Euphrates flood, controlled by rain and snowfall in Turkey, arrives too late in the parched southern summer to be of any use for watering crops. Before 4000 B.C.E., the growing season was later and longer, so the arrival of the flood often coincided with the time when water was most needed.

Each community flourished amid a mosaic of local environments that included natural flood basins, marshes, and levee areas with fertile soils that were subject to irregular flooding. This palimpsest of different environments provided an important safety net, to the point that if crops failed, the villagers could become virtually full-time herders or could fall back on fish, waterfowl, and other marsh foods.

Satellite images hint that many early farming communities near the Euphrates may have lain on natural rises in the plain amid marshes. In many cases, inhabitants may have watered their fields with the help of small canals or channels linked to breaks in nearby levees. These breaches were a natural way of leading water onto the floodplain, to be steered by local communities toward their fields, though such practices carried the risk of much larger breaches caused by unusually high floods that would produce uncontrollably large quantities of water. The fields watered by these simple diversions produced crops of barley and wheat. Irrigation resulted in higher crop yields per acre, even if the physical effort involved was greater. The increased yields also came from genetic changes in barley, which has two rows of grains in its uncultivated form. Irrigating the seed led to selection in favor of a six-row variety, which rapidly came into use throughout the Near East. Six-row barley produced at least 50 percent more grain, one of several factors that led to much higher crop yields—and to people living much closer together and subsisting off less land than when they relied on rainfall alone.

All this farming activity was on a small scale, based on ancient methods that relied on receding floodwaters and simple gravity canals, with very little systematic investment on the part of the villagers. Decisions about canal maintenance and diverting water to individual fields were made within the community, perhaps on the basis of kin ties—we can only speculate. These were centuries, too, when rainfall was greater than today, when a combination of summer monsoon rains and higher flood levels would have inundated large areas of the poorly drained southern plains for months on end. As long as spring and summer rainfall remained plentiful, small farming villages and nomadic herders could support themselves comfortably, with plenty of grazing and easily irrigated land to go around. Then, in about 3800 B.C.E., Mesopotamia became significantly drier, with momentous consequences for those who irrigated its soils.

CHAPTER 7

The Lands of Enki

"WHEN YOU ARE ABOUT TO TAKE hold of your field, keep a sharp eye on the opening of the dikes, ditches, and mounds [so that] when you flood the field the water will not rise too high in it. When you have emptied it of water, watch the field's water-soaked ground that it stay virile for you."[1] A farmer's almanac of the third millennium B.C.E. from the Sumerian city of Nippur allows us to look over the shoulder of a villager as he prepares to water his land. "In days of yore a farmer instructed his son," begins the manual. The 107 lines that follow once served as a pedagogue's manual and distill the experience of centuries, passed by word of mouth from one generation to the next. (Sumer, in southern Mesopotamia, was the heartland of the Sumerian civilization.)

The cumulative knowledge of centuries served village farmers well. Every landowner knew the local topography intimately—the subtle contours in the flat landscape, the places where floodwaters breached levees even in drier years, and the locations of the most fertile soils. Over many generations, village elders became expert at knowing when to plant their emmer wheat and barley, at identifying the strategic moment when frost would no longer menace growing seedlings. Judging from clay tablets compiled in later times, they also learned the telltale signs of potentially catastrophic floods and impending low-water years. All kinds of arcane information passed from father to son by word of mouth, and, eventually, in tablet form. Years of hard-won experience and good rainfall produced much higher crop yields than those obtained from dry agriculture.[2]

As the centuries passed and population densities increased, the small hamlets of earlier times became clusters of rural communities located around a larger settlement. By 5200 B.C.E., five centuries after the first known colonization of southern Mesopotamia, the largest of these population clusters was a town housing between two thousand and four thousand people and covering about twenty-five acres (ten hectares). Such communities were the exception rather than the rule. Life still revolved around the household and ancient ties of kin, which sustained small, dispersed villages, just as they had since the earliest days of farming. Household and kin were the glue that provided the labor to clean natural channels, to divert water into fields. The few larger villages and small towns were focal points for trade, for decision making, and for ritual activities. It was there, at places like Eridu, close to the Euphrates, that the first temples appeared, dedicated to ancient patron gods and to the ancestors.

Sumerian legends from later centuries tell us that Eridu was the oldest town of all, the dwelling place of the god Enki, lord of the abyss and god of wisdom. "All lands were the sea, then Eridu was made," proclaimed a long-established creation legend inscribed on a clay tablet many centuries later. Archaeology adds legitimacy to legendary claims. Archaeologists Fuad Safar and Seton Lloyd deciphered layer after layer of barely visible mud brick in Eridu's temple mound. They used digging techniques developed by German scholars at Babylon before World War I, which employed picks to "feel" different soil textures and identify courses of unbaked bricks. Adding brushes and compressed-air hoses to their armory, Safar and Lloyd exposed a solid temple platform that had been extended again and again to add new shrines. In the end, they uncovered sixteen temples on the same site, the original a small mud-brick shrine erected on clean sand and dating to 5500 B.C.E. or even earlier. The skeleton of a sea bass, caught in brackish waters near the town, still lay on the offering table. Five hundred years later, a magnificent stepped ziggurat stood in the center of what was now a busy city, its façade adorned with brightly colored bricks, surrounded by a sacred enclosure at least 200 yards (180 meters) square. Residential quarters and markets pressed on Eridu's sacred precincts.[3]

Eridu and Uruk, another burgeoning urban community upstream, came into being at a time when Mesopotamian society was changing profoundly in the face of much more challenging environmental conditions (map: figure 6.1).

WHY DID HUNDREDS of small farming villages rapidly coalesce into crowded cities? This question has fascinated generations of scholars, who have invoked every conceivable factor possible. A need for defense, the appearance of new, authoritative rulers, a growing population of non-farmers triggering food shortages and closer control of irrigation works, a great expansion of international trade—there are almost as many potential causes in the academic literature as there are archaeologists. What actually happened will never be known for sure, but most experts now agree that a combination of climatic change, rapidly expanding long-distance trade, and the growing power of temples and their leaders was among the tidal streams that triggered major social change. There is only one certainty in this complex equation. Without high-yield irrigation and ample water supplies, cities and civilization would never have developed in the land between the rivers.

A trend toward larger, more densely populated towns and villages was well under way after 5000 B.C.E., but one of the catalysts for more rapid change occurred in about 3800 B.C.E., when the Near East and the eastern Mediterranean region became significantly drier for well over a thousand years.[4] Solar insolation, the rate of incoming light at the earth's surface, declined, a trend documented by radiocarbon dates from tree rings and lake cores across the world. The changes resulted from shifts in the earth's angle to the sun. Almost immediately, the southwestern monsoon that brought summer rainfall from the Indian Ocean to Mesopotamia weakened and shifted to the south. The rains faltered, began later, and ended much earlier. Now the summer flood arrived after the harvest, which meant that the near-ripe crops received much less water. At the same time, the summer inundations were far smaller than before, for rainfall and snowfall had also declined in Turkey, the source of the Euphrates and the Tigris. The climate became increasingly unstable, with prolonged drought cycles that played havoc in small communities tied to capricious river channels and diverted water.

Irrigation farming had never been easy, even with a longer rainy season. Dozens of small settlements lay along meandering stream tributaries, such as are commonplace on any alluvial plain. Although we lack archaeological proof, it seems likely that the farmers who fed the growing cities still cultivated narrow bands of fields along natural levee back slopes and the margins of seasonally flooded depressions. Perhaps some larger settlements lay close to brush weirs that serviced small canals and nearby fields. But there was no central authority. As archaeologist Robert Adams, a veteran of irrigation surveys, and others have pointed out, a possible analogy comes from descriptions by nineteenth-century travelers, who wrote of a patchwork of small tribes, all with their own irrigation works that varied in extent and shifted constantly in response to rapidly changing conditions.[5] There was no way that anyone could invest in elaborate, permanent irrigation works in a world where stream channels shifted constantly and locally. Chronic water shortages were an endemic reality of daily life. Each tribal group was diverse in its makeup, with farmers and herders part of the community, so that decisions about shifting over to more herding and vice versa unfolded without conscious effort. There were no despotic leaders, no powerful rulers to dictate policy or allocate water or repair channels. Power lay in the hands of tribal sheiks, whose authority depended on their personal qualities, as well as traditional loyalties and reciprocal ties that linked every member of society.

As drier conditions settled on the south, the villagers fell back on the traditional safety nets of their herds and the bounty of the marshes. But there were limits to the population that could be supported by such resources in a changed world where many people now lived in much larger, more densely populated communities. The towns were victims of their own success, of prosperous centuries spent diverting water from seemingly limitless rivers. The major feeder channels were like branching trees with progressively smaller boughs, which eventually became mere twigs feeding small amounts of water to remoter fields (figure 7.1). Obviously, the strategic points were the nodes where the main channels and levee breaches branched off from the main river, for it was here that people could control what water reached whom and when. Such control

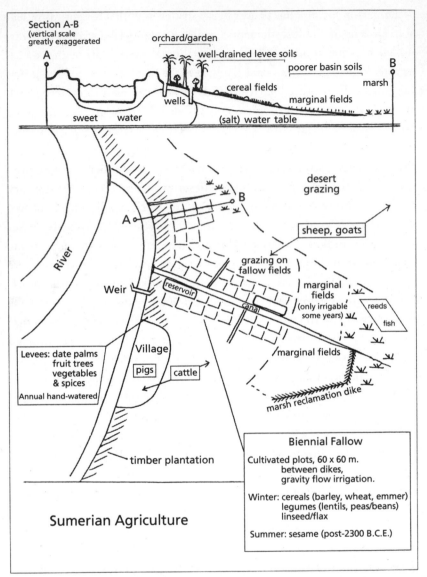

Section A-B
(vertical scale greatly exaggerated

orchard/garden

well-drained levee soils

poorer basin soils

A

cereal fields

marsh

wells

marginal fields

B

sweet   water

(salt) water table

desert grazing

B

A

sheep, goats

grazing on fallow fields

River

marginal fields

Weir

reservoir

(only irrigable some years)

reeds

canal

fish

Levees: date palms
fruit trees
vegetables
& spices

Village

marginal fields

pigs

cattle

Annual hand-watered

marsh reclamation dike

timber plantation

Biennial Fallow

Cultivated plots, 60 x 60 m.
between dikes,
gravity flow irrigation.

Sumerian Agriculture

Winter: cereals (barley, wheat, emmer)
legumes (lentils, peas/beans)
linseed/flax

Summer: sesame (post-2300 B.C.E.)

Figure 7.1 *A hypothetical Sumerian irrigation system. (From Nicholas Postgate,* Early Mesopotamia, *fig. 9.1. Copyright © 1994 by Nicholas Postgate, Routledge. Reprinted by permission of Taylor & Francis Books, UK)*

assumed ever-greater importance in a time when rainfall had declined and flood levels were on the decline. The growth of Uruk, close to the Euphrates, is a barometer of the far-reaching changes that now gripped Mesopotamia.

URUK WAS THE home of the mythic Gilgamesh, hero of the epic that bears his name.[6] It lies east of the present-day Euphrates, on the banks of the now-dry but ancient Nil channel, which once provided not only water for irrigation but also access to trade routes up- and downstream. (In its heyday, the city flourished southwest of the river, which has now shifted further to the east.) Canals separated Uruk's growing neighborhoods, so much so that some imaginative modern observers have called it the "Venice of Mesopotamia." The future city came into being when two large farming villages coalesced into a single settlement, sometime around 5,000 B.C.E. Serious growth began a thousand years later, when two major shrines formed two quarters of what was now a near-city. Crowded neighborhoods formed of closely packed houses with courtyards surrounded the shrines, the major one, associated with Gilgamesh, dedicated to the goddess Inanna, the deity of love, who was said to have planted a willow tree taken from the waters of the Euphrates at the future site of her temple.

By 3500 B.C.E., Uruk was much more than a large town. Satellite villages, each with its own irrigation system, extended as much as six miles (ten kilometers) in all directions. Some were predominantly fishing settlements; others specialized in pottery manufacture or metal artifacts. A close web of interdependencies linked village to village and villages to the city, with its bustling markets. Not only was the growing city a major religious center, but its temples and bazaars served as a major trading center that linked Uruk with a far wider world. Much of this trade moved by water up and down the Tigris and the Euphrates, for the city lacked timber and metal ores, while its verdant landscapes and numerous potteries supplied foodstuffs and clay vessels to merchants far away. As the city expanded, so did the numbers of non–food producers, whose sustenance came from the hinterland. Uruk also provided protection

Figure 7.2 *A scene of complete aridity. The site of Uruk today. (Nick Wheeler/Corbis)*

from covetous neighbors, who craved its water supplies and abundant commodities.

This nascent and rapidly spreading metropolis was one of several prototypes for the flourishing Sumerian cities that were to appear in the south during the fourth millennium B.C.E. The trend toward city living throughout much of Sumer was pervasive. By the end of that millennium, over 80 percent of the population of the south lived in settlements covering at least twenty-five acres (ten hectares), the first truly dense urban living in history.

As Uruk grew, a spurt of rapid technological innovation transformed life in the city and also in the fields, where agricultural production soared to feed thousands of non-farmers. First came a shift along the southern reaches of the Euphrates from basin irrigation, which had been near-universal in earlier times, to new methods based on furrows. Basin irrigation submerged entire fields under a shallow layer of water, which soaked rapidly into the soil. The fields were usually square and, of necessity, perfectly level; otherwise the submergence would be uneven. A single family managed each field, which worked well in a village environment, where the local channels could be cleared and straightened on a piecemeal basis as the population grew or the need arose. Furrow irrigation was far more standardized and involved long strip fields, laid out in parallel close to one another. Every field had a slight, regular slope, the upper end lying close to the canal where the water came from, the lower portion leading to a drainage basin or a marsh. Water only inundated the furrows, irrigating the field through horizontal, rather than vertical, percolation. The farmer's almanac quoted earlier gave explicit instructions: "When the barley has filled the narrow bottom of the furrow, water the top seed. When the barley stands up high as [the straw of] a mat in the middle of a boat, water it [a second time]."[7] Strip-field systems required far-from-casual organization, for they had to be laid out with great care to maximize the available terrain and also to position them relative to the raised banks of the canal. Such systems could come into being only with very close coordination between those who cultivated them, and under some form of carefully planned management. They formed large blocks of fields laid out in herringbone patterns off both sides of a canal.[8]

No one knows when strip fields and furrow irrigation came into use, but early tablets from Uruk speak of such methods as early as 3100 B.C.E., and it was almost certainly much earlier than that. The same texts tell us that the fields were organized in large blocks and were centrally managed, probably by temples.

Long strip fields are almost impossible to dig by hand while keeping the furrows straight. Nor was it feasible to place the seed in the ground by hand. Enter the scratch plow, commonly known as an ard—basically, a hooked digging stick, originally made of wood, then later fitted with stone and, ultimately, bronze and iron plowshares. We don't know when plows first came into use, for we have no actual specimens, only crude depictions of them in the form of pictograms. Two or three oxen pulled the plow, beasts that were far from easy to manage and steer straight, quite apart from the costs of feeding them. However, plows were ideal for grooving the soil with shallow furrows along the lengths of long fields, where far fewer turns were needed than in a square plot. They saved enormous amounts of time when compared with hoe cultivation. When the field was ready for planting, the farmer used another innovation. He fitted a funnel onto the plow to turn it into a seeder that dropped seeds deep into the furrows. Seeder plows, known from a pictographic sign, *apin*, were in wide use between 3400 and 3100 B.C.E. They reduced seed wastage and increased crop yields by as much as 50 percent.

The combination of the strip field, the ox-drawn plow, and channel irrigation created an economic powerhouse behind the numerous, and often volatile, Sumerian city-states that emerged after 3100 B.C.E. The innovations came at a time when the rising sea levels of the Persian Gulf had reduced gradients and increased sedimentation. The levees along rivers and channels brought water to the fields at a slightly higher elevation. A combination of animal traction and much more efficient, higher-yielding irrigation systems revolutionized agriculture in the south. The village farming of earlier centuries gave way in many places to what one can only call a simple form of mechanized farming, the ancient mode of the farm worked by family and kin passing into history there as productivity increased, perhaps in some places by 500 to 1,000 percent.

Uruk and later city economies were based firmly on barley and sheep, the latter providing wool for textiles. Barley had two benefits for farmers in the region: It matures very rapidly and is tolerant of saline soils. With the climate changes caused by the shift in monsoon rains, the grain acquired a decisive advantage in the south, where floods could threaten crops as they approached harvest in late spring. Early reaping also avoided another hazard: locusts, which could strip a field in minutes. And salinity was a constant problem in this area of very slight gradients and higher water tables. It was no coincidence that barley covered about 90 percent of the cultivated land in the south in 2500 B.C.E., with wheat becoming more significant as one moved upstream. Barley, grown in strip fields, formed the food surpluses that fed southern cities. Much of the harvest went to the temple, after a deduction for the workers' use, this over and above the corvée labor expected of them during specific periods such as the reaping season.

Corvée labor (the word ultimately comes from the Roman word *corrogare*, "to requisition") was unpaid labor imposed on commoners by their social superiors. It differed from slavery in that the laborers were free citizens and, in Mesopotamia, it involved set periods of work, for canal maintenance, harvesting, and so on. The temples lay on mud-brick platforms, which raised them above floodwaters and gave them prominence amid a flat landscape. As they crumbled, they were rebuilt, rising in stages, like the famous ziggurat at Ur, and lying within walled enclosures. Sumerian temples were shrines and administrative centers, storehouses and trading centers. Their priests administered lands owned by the gods, receiving most of the produce from them as tribute, as well as the labor to produce it, thought to be a gift to the deities.

Each long field was of a standard size of about seventy-four acres (thirty hectares), or one hundred *iku* in Sumerian parlance, and was cultivated by a single *engar*, a farmer or agricultural manager, equipped with a plow and two or three pairs of oxen. When more people were needed, corvée provided them, men drawn from the extended families anchored to the land. The seasonal workers were fed; all other costs, such as seed for the next year, representing about a third of the harvest, lay with the villages, which had to feed themselves. According to later texts, administrators calculated the potential yield before reaping took place, then took a flat amount of it before the harvest encouraged stealing and other forms of fraud.

Easily stored and handled, barley was the foundation of an economic system that linked the temple (and, later, royal palaces), the village, and temple (and royal) lands—but only after the investment of large amounts of human labor. Other agricultural products, such as dates, lettuces, and onions, rotted rapidly. Everyone grew their own supplies of these without restriction.

WITHOUT A DRAMATIC increase in productivity and the huge grain surpluses associated with it, cities like Uruk would never have come into being.[9] But the process of change involved far more than a direct link between increased productivity, larger food surpluses, and population growth. The strip-field systems and their densely packed furrows required a level of supervision that transcended the family and the kin group. A new social element entered the picture: some form of central authority that supervised irrigation and farming on a considerably larger scale. This was the mechanism that ensured a degree of standardization in both agriculture and irrigation, something very different from the more informal and volatile village irrigation systems of the past. One such instrument of supervision in later times was the *gugallum*, or "canal inspector," who may also have been responsible for canal cleaning and maintenance. Despite their titles, such officials are rarely mentioned in tablet records. Almost certainly, they were local people, whose responsibilities were confined to community irrigation systems, where they worked closely with their fellow villagers.

It's easy to speak of food surpluses and then envisage small regiments of minor officials who inventoried the harvest and took a portion of it as a tax in kind. In fact, the ancient texts mention no such taxation. Instead, the taxation came in a form of the corvée assessment that provided the labor for irrigation works and improvements to the agricultural infrastructure, as well as for building temples and other public works, such as city walls. Growing settlements like Eridu and Uruk functioned on a system of carefully issued rations for corvée workers as well as non-farmers. The crops grown on temple and palace farms supported artisans, specialists of all kinds, and officials, as well as the ruler and the priests. Everything flowed to the center, the labor being devoted to the common good and to the honor of the deities who presided over the city and their representatives. The austere, basically simple life of the village farmer began to change early on, as shrines and temples appeared in larger settlements. By the time the enormous temple of Eanna rose in the heart of Uruk, around 3100 B.C.E., the city covered nearly 550 acres (200 hectares), with a population of between fifty thousand and eighty thousand people crowded within its walls. Temples like those in Uruk became potent economic, political, and social forces, the places where elaborate public ceremonies communicated powerful myths, creation legends, and religious ideologies. Such organizations were the catalysts for a sense of community, for a subtle coercion that had farmers willingly giving labor for the common good—and feeding themselves along the way.

For all the ideological coercion, those who fed the city remained in the nearby countryside, creating a bipartite world of cities and villages. The villages produced food; the cities were centers of exchange, manufacturing, and administration, of religious ceremony. A third-millennium proverb aptly remarks that "the outer villages maintain the central city." A hinterland of villages extended as much as six miles (ten kilometers) out from the urban core. In time, however, the rapid expansion of Uruk and other settlements, like Eridu, Umma, and Ur, submerged many of the lesser communities nearby in a way that did not occur further north.

The Italian scholar Mario Liverani emphasizes what a big step it was for irrigation farmers to move away from the strategies of self-sufficiency and self-reproduction inherent in subsistence farming and toward an existence that forced them to contribute to a much larger, more impersonal authority, through either physical or ideological coercion.[10] The farmers were the "outer circle" of the newly complex urban society, providing labor without receiving much in return, except the satisfaction of feeding themselves and laboring for Enlil or whatever patron god presided over the nearby city. The labor, and the food surpluses, of the outer circle fed the inner one: artisans, officials, and priests, who served the temple and received rations in exchange. This unequal system of redistribution supported not an egalitarian society but one where social inequality and privilege had become foundations of new city-states. A large body of what is called "wisdom literature" reinforced the notion of an elite and commoners who depended on one another. A proverb stresses cooperation and the need for hierarchy in society: "Don't drive out the powerful, don't destroy the wall of defense." Some early Sumerian literature sets down creation legends that are firmly based in the soil, such as that of Ninurta, the son of Enlil, who averted drought and famine caused by a Tigris flood:

> *The high water it pours over the fields,*
> *Behold, now, everything on earth,*
> *Rejoiced afar at Ninurta, king of the land,*
> *The fields produced abundant grain,*
> *The vineyard and orchard bore their fruit,*
> *The harvest was heaped up in granaries and hills.*[11]

The image of gods and rulers carrying hoes and baskets was powerful in Sumerian ideology. So was the vision of the southern Mesopotamian landscape hedged in by two great irrigation channels—the Tigris and the Euphrates—and hemmed in by mountain ranges that served as embankments against disastrous floods. Thus, the burdensome labor of canal digging and agriculture acquired a value equal to that of other tasks in an unrelenting environment of startling, and often under-recognized, diversity.

IRRIGATION STILL INVOLVED the management of natural channels, abandoned river meanders, and new drainages created by avulsion events. Cuneiform records on clay tablets tell of canal cleaning during the third millennium B.C.E. The number of natural channels fell, but the belt of irrigated cultivation grew broader. As archaeologist Tony Wilkinson points out, in general the few major watercourses and branch canals that extended the limits of cultivation in some areas determined the landscape. Another archaeologist, Hans Nissen, writes aptly of irrigation oases, "threaded like pearls along the main water courses."[12] As human water management intensified, natural channels from major watercourses became straighter and more artificial, the straightening being accomplished as much to allow boats to pass easily as to increase water for agriculture. Most of this work was carried out piecemeal without any form of centralized planning on the part of growing cities, which had now changed the character of human society in fundamental ways.

By 3100 B.C.E., the southern cities had become a jigsaw of intensely competitive city-states, each with its own hinterland of villages and irrigated lands. Sumerian civilization was born. Now the nodes of Mesopotamia's water supplies fell under the rule of secular leaders, known as *ens* or *lugals*, who presided over trade and war, agriculture and diplomacy. An intricate and ever-changing maze of political alliances and individual obligations of friendship had linked community with community since the beginnings of farming. These volatile alliances now operated on a much larger scale, as political, economic, and spiritual power passed into ever-fewer hands.

Understanding the changes in irrigation and water management that accompanied these developments is a complex archaeological task, involving the use of ground surveys, geomorphology, and cuneiform-tablet records. By mapping the remains of levees from old branches of the two great rivers, researchers have been able to produce a topographic framework for mapping later channels. At some point, bifurcating Euphrates channels in the north extended far eastward, so much so that they may have joined a former course of the Tigris, forming a channel filled with water from both rivers. No channel was permanent; they changed repeatedly, like one that saw the Euphrates shift its course abruptly from east of the nascent city of Nippur to much further west. On another occasion, sometime between 3000 and 2700 B.C.E., the Euphrates shifted from a line through Nippur, Shuruppak, and Uruk to a more eastern course that favored another city, Umma.

The main settlement pattern followed the elongated levees that bounded the major channels. Many smaller settlements, some little more than hamlets of reed-and-mud houses now buried under later alluvium, may have surrounded cities like Umma. Their local irrigation systems would have been nourished by small watercourses leading off each bank of the main river or by new channels dug from the river at a point well upstream and running roughly parallel to the main channel. In general, however, the overall layout of the fields depended on the gradient and the configurations of nearby water channels. There was little standardization—the dictates of gradient irrigation and the relatively limited numbers of people available to work on canals and other waterworks made this impossible.

The same realities could dictate the layout of entire cities. Nippur lay on the left bank of the Euphrates, which flanked the western side of the city, while a large canal bisected the teeming settlement. Another canal lay to the east. Water channels for both agriculture and boat transport divided different quarters of the city, which now survive as sandy mounds. The Sumerian city of Larsa, south of Uruk, used water channels to separate the administrative, religious, and residential areas of the settlement. *The Epic of Gilgamesh* tells us that Uruk was subdivided into three 1-square-mile (2.6-square-kilometer) segments. One quarter was the city; a second, gardens; a third, brick pits. A final .5-square-mile (1.3-square-kilometer) quarter belonged to the goddess Ishtar's temple.

IT WAS NOT until the first millennium B.C.E. that irrigation technology grew more sophisticated and became capable of manipulating water by

reconfiguring the landscape on a considerably larger scale. In the 1840s
B.C.E., Sin-iddinam, the ruler of Larsa, boasted how "in order to provide
sweet water for the cities of my country . . . [An and Enlil] commissioned me to excavate the Tigris [and] to restore it [to its original bed]."[13]
Deliberate waterworks could militate against the natural processes of
avulsion. However, artificial canals designed to bring water to dry areas
or to create shortcuts could conceivably lead to catastrophic levee
breaks when the canals overflowed at times of heavy flood.

Who labored over these expanding irrigation works, which required
large numbers of human hands? Yale University archaeologist Frank
Hole believes that the hands belonged to "the landless and destitute,"
those who had fled their traditional villages when rains had failed and
irrigated land had become dry. To undertake large-scale canal digging
would have placed a severe burden on the existing labor pool, which was
fed by government rations and also built temples, city walls, and other
public works.

Some idea of the labor involved in building longer canals comes from
information in Old Babylonian mathematical texts and administrative
tablets from Ur, dating to about 2100 B.C.E. Ration budgets tell us that a
canal 31 miles (50 kilometers) long, 6.5 feet (2 meters) deep, and between
14 and 23 feet (4.3 and 7 meters) wide would have taken 1,925 men sixty
days to dredge and between 1,500 and 4,100 men sixty days to excavate.
Longer, more ambitious canals would have consumed much more people-
power and time. Whatever the scale of the canal digging, it would have
tied up the agricultural labor force for months on end and reduced crop
yields to dangerously low levels. The only way to construct large-scale
irrigation works was to use imported labor such as the landless and destitute or, increasingly, enslaved prisoners of war, a practice that was
commonplace in later times (see chapter 8).

Until such labor sources were available, the only solution other than
digging the occasional large canal was to manage natural avulsions in
such a way that new channels would remain open and expand potentially
irrigable acreage. The new channels would also serve as a safety valve
for carrying away water in times of exceptional flood. Irrigation, then,
was local, its scale limited by the vagaries of poorly drained topography
and the volatility of the river regimens. These changing river systems and

the concentration of settlement on major nodes, especially of the Euphrates, gave ambitious rulers numerous opportunities to provide water for their subject communities, and to withhold it from their neighbors. Southern communities were poised between opportunities for rapid growth presented by channel shifts, especially if cleverly managed, and slow starvation resulting from diversion of water during political conflicts. Then there were always the sudden natural catastrophes that could wipe out entire irrigation systems in a few hours.

After about 2700 B.C.E., walls appeared around many cities, among them Kish in the north and Nippur to the south. The Sumerian world pitted city against city in violent disputes over land and water. The most powerful cities in the south were Lagash, Ur (one of Sumer's most ancient cities), and Uruk. Volatile diplomatic relationships and quarrels over trade were commonplace: "Be it known that your city will be completely destroyed! Surrender!"[14] The rhetoric of aggression was as shrill in 2600 B.C.E. as it is today. Few of these raids, spats, and all-out wars have come down to us. One that has involved a long-forgotten boundary dispute when Mesalim, the highly respected ruler of Kish, far upstream, mediated between Lagash and Umma over a strip of land known as Gu'edena, on the edge of the floodplain. Urluma, ruler of Umma, promptly expropriated the land. Eannatum, king of Lagash, went to war on two occasions to recover it. He prevailed, forcing Umma's ruler to swear humiliating oaths. In fact, Umma may have leased part of the strip in return for rent and produce, then failed to pay up. There was always an excuse for threatening oratory. A later Umma ruler "diverted water into the boundary channel of Ningirsu and the boundary channel of Nanshe. He set fire to the monuments." Lagash prevailed and rebuilt the levee in expensive stone to the accompaniment of harsh threats against Umma's lord: "May Ningirsu, after casting his great battle net upon him, bring down upon him his great hands and feet. May the people of his city, having risen up against him, kill him there within his city!"[15] The fragmentary records of the long-festering dispute talk of raising "the battle net of Enlil," for the battles were always fought in the name of the gods. Inevitably, the victor "set up burial mounds."

One could write glibly about landscapes of power manipulated by kings, but reality was much more complicated because of the shifting configurations of the natural landscape, many the consequence of human modification of what nature had fashioned. The history of the Sumerians was one of cycles of boom and bust, of population rises and declines, caused not only by changing landscapes but also by the gradual rise in salinization, which affected agricultural productivity. At Ur, for example, by 2000 B.C.E. crop yields had fallen by half from those of earlier times, despite the at least six major branches of the Euphrates that supplied water for the city.[16] The villain may have been rising salt levels. Denser populations, and insatiable demands for more food, had led to year-round farming—and the shortening of critical fallow periods that had earlier helped leach salt from the soil.

The city-states of the south were the product of a long-term method of feeding people triggered by increasing aridity, part of a unique way of responding to environmental crisis. Long-distance trade also played an important role in state formation. Cities like Ur forged networks of trading contacts that extended far up the great rivers into Turkey and as far west as the Mediterranean coast. By 2500 B.C.E., Sumerian lords were competing with growing cities far north of southern Mesopotamia and as far away as northwestern Syria. They attacked trade routes and annexed neighbors, but were often distracted by internecine strife and petty rivalries close to home. Inevitably, broader territorial ambitions came into play. The Akkadian cities to the north of Sumer eventually turned on their southern neighbors. In 2334 B.C.E., King Sargon of Agade, south of Babylon, defeated a coalition of Sumerian city-states led by King Lugalzagesi of Ur. He smote Eridu and brought Lugalzagesi in a neck stock to the gates of Nippur. Sargon crafted a huge empire that eventually encompassed all of Mesopotamia and lands far to the east and west. However, Sargon's much larger domains were far more vulnerable to increasing drought than their smaller predecessors and soon collapsed in a familiar pattern of volatility, sometimes in the face of climate change that had long plagued early states in Mesopotamia.

Drought rippled across Akkadian domains with catastrophic effects. We know this from Tell Leilan, on the Habur Plain west of the Euphrates, an imposing city with a citadel that lay in the heart of a tightly knit, closely organized agricultural landscape when the Akkadians conquered it in 2300 B.C.E. For a century, the city prospered, as hundreds labored on public projects, including large-scale waterworks. These bought some protection against year-to-year fluctuations in river floods—as long as there was enough rainfall to sustain average inundations in the Euphrates. But in 2200 B.C.E., a major volcanic eruption to the north coincided with a 278-year-long drought that turned the Habur into a dust bowl. Tell Leilan became a ghost town. Elsewhere, thousands of herders found their summer pastures on the plains devastated. In response, they moved southward onto settled farming lands, jostling their way onto cultivated plots. The ruler of Ur was so alarmed by the threat that he erected a 112-mile (180-kilometer) wall named the "Repeller of Amorites" to keep herder immigration in check. His efforts were to no avail. Ur's hinterland suffered a threefold increase in population as fruit trees died and the authorities frantically straightened irrigation channels to maximize much-reduced water flow. Cuneiform tablets tell us that Ur's agricultural economy collapsed. Officials were reduced to distributing grain in minute rations.

Mesopotamian civilization endured to thrive anew, once rainfall returned to its previous seasonality. People repopulated the Habur and Assyria; Tell Leilan prospered once more; the ideologies and institutions of older times survived to become the blueprint for the great empires that rose on the foundations of earlier city-states. And with the new empires, irrigation agriculture became a state business—even if its roots remained in the hands of village farmers managing water and crops as they had done since time immemorial—just like it did among the ancient Egyptians of the Nile Valley.

CHAPTER 8

"I Caused a Canal to Be Cut"

MESOPOTAMIA WAS A WORLD of climatic extremes and violent floods, of torrential rains and steaming-hot summers, a place where political rivalries always simmered. Egypt, a unified state after 3100 B.C.E., known to its citizens as *Kmt*, the "black land," was a linear kingdom, watered each year by the rising floodwaters of the Nile. The contrast was dramatic. In Egypt, the pharaohs presided over a seemingly unchanging world. Each morning, the sun rose in a cloudless sky and, in its personification of the god Re, journeyed like a solar bark across the heavens. Life seemed to be easy here, and often it was. That inveterate Greek traveler Herodotus wrote of the Egyptians, "They gather their crops with less effort than anyone else in the world . . . They do not work at breaking the land up into furrows with a plough, they do not have to wield hoes . . . Instead, the river rises of its own accord and irrigates their fields, and when the water has receded again, each of them sows seed in his own field and sends pigs into it to tread the seed down."[1] After that, all they had to do was wait for the crop to ripen, and then the pigs threshed the grain—or so Herodotus claimed. We have lived with a myth of effortless ancient Egyptian agriculture ever since, despite abundant evidence of a capricious, unpredictable Nile, of years when the inundation faltered, bringing famine. Just as dangerous were exceptionally high floods that swept entire villages and centuries of irrigation works before them. Despite these realities, the Egyptian relationship with the Nile remained basically unchanged for more than three thousand years.

Each summer, heavy monsoon rains in the Ethiopian highlands swelled the waters of the Blue Nile and Atbara rivers far upstream. (Most of Egypt's water comes from the Blue Nile, which starts at Lake Tana, in Ethiopia. The White Nile rises in the central African lakes. The two rivers meet at Khartoum, in Sudan.) The silt-heavy flood surged northward, reaching its height over about six weeks between July and September.[2] In Egypt, the Nile rose until it overflowed its banks under cloudless skies. During the season of Akhet, the waters of the inundation spread over the floodplain, which sloped away from the main channel. This was the time of anticipation. A pyramid text observes, "They tremble that behold the Nile in full flood. The fields laugh and the river banks are overflowed. The god's offering descends, the visage of the people is bright, and the heart of the god rejoices."[3] As the current slowed, the river deposited its silt, then gradually receded. The Nile fertilized and watered the Egyptians' fields, replenished marshlands and lush meadows. In a good flood year, the river valley became a vast shallow lake. Villages and towns became islands. As the waters receded, in the season known as Peret, the farmers planted their winter crops, which thrived on the damp soil and the high water table. A combination of capillary and hygroscopic soil moisture and the high water table allowed crops like barley and wheat to grow and mature during the completely dry winter months. Each year, the inundation brought life to the fields and food to a long-lived, intensely conservative civilization.

The gradient of the Nile is so shallow that radiating canal networks like those used in Mesopotamia were impracticable. Almost everywhere, the irregular meanderings of the river channel and sudden topographic changes made any form of irrigation other than simple flood-basin works impossible. Each village took care to conserve and direct the floodwater. Entire communities would work on natural levees, making them higher and stronger, and effectively artificial. They enlarged and dredged natural overflow channels, blocked off or joined natural drainage channels with earthen dams and sluice gates, and subdivided flood basins. They also took steps to control water access and retention in different subdivisions of the basins by using temporary cuts in levees or a network of short canals and masonry gates. The system was inadequate to cope with exceptionally high floods or deficient ones. But it fed the 1.25

Figure 8.1 *Egyptian farmers sowing seed. The tomb of Nakht, a scribe and priest under pharaoh Tuthmosis IV (1419–1386 B.C.E.). (©British Library/HIP/François Guenet/Art Resource NY)*

million people living in Egypt during Old Kingdom times (2575–2180 B.C.E.) and the 2 million dwelling there during the New Kingdom, between 1550 and 1070 B.C.E., albeit with occasional famine years.

Ancient Egypt glittered with wealth and the brilliance of the pharaoh's court, but remained a predominantly agrarian society. Food came from villages, from the labor of commoners. Their lives were remote from the realm of the kings, who taxed them in labor and produce. The ideas and ideals that legitimized the rule of the pharaohs stemmed from a carefully crafted ideology that depicted a great king, a terrestrial ruler, who symbolized the triumph of order over universal chaos. Their power came from this philosophy and also from their skillful administration of people and their control of agricultural surpluses and labor. The pharaohs manipulated these surpluses for the benefit of a tiny elite, as well as to pay many members of the general population in rations from the state's granaries. The early rulers developed a centralized bureaucracy that directed labor and collected taxes in produce and other commodities. In Egypt, the terms "father," "king," and "god" were inter phors for one another and for a form of political power based on inequality, considered part of the natural order established by the gods at the time of creation. Pharaonic rule worked so well that it survived the ups and downs of political history for three thousand years.

The pharaoh had power over the Nile flood, over rainfall, and over all people, including foreigners in his domains. He was a god, respected by everyone, a tangible divinity who was the personification of *ma'at*, best translated as "rightness." *Ma'at* was far more than rightness, though. It was a seemly order and peace. A massive hereditary bureaucracy effectively ruled Egypt. An army of officials supervised tax collection, the harvest, and the administration of irrigation. Only a tiny number of Egyptians were literate. Then as now, information was power, and to become a scribe was to join the elite. "Be a scribe. Your limbs will be sleek, your hands will grow soft. You will go forth in white clothes, honored, with courtiers saluting you," a young man was advised.[4] Scribes supervised every moment of the food cycle. They measured grain yields on the threshing floor, monitored the grain in transit to curb pilfering, and inventoried granaries—even the loaves turned out by bakers.

There were never jostling city-states along the Nile, only a unified kingdom where the pharaohs presided over a realm of predictable cycles—from the rising and setting of the sun day after day to the Nile inundation. In Egypt, remarked the Roman author Pliny the Elder, "the Nile plays the part of farmer." Its rulers knew the effectiveness of flood irrigation only too well. Their wealth made Egypt a tempting quarry for ambitious conquerors. In 671 B.C.E., Esarhaddon, king of the Assyrians, crossed the Sinai Desert and invaded the Nile Valley. He captured the pharaoh's capital at Memphis and most of the royal family. "Egypt was sacked and its gods were abducted," claimed the grandiloquent monarch. He installed Assyrian governors, but they soon fled in the face of a restive populace. Esarhaddon's vengeful son, Assurbanipal, descended on the Nile four years later and repossessed Memphis. His soldiers ransacked the temples of the sun god Amun at Karnak and Luxor at Thebes, emptied the royal treasury, and devastated the holy city. The pharaoh fled and never returned. His predecessors had long dabbled in the politics and wars of the wider eastern Mediterranean world, but their domains had always remained sacrosanct. The Assyrians did not last long, but they opened the door for other conquerors and for new ideas and inventions from harsher environments, including some for water management. Centuries passed before later invaders turned Egypt from a self-sufficient state into an international granary. Even then, almost all agriculture was decentralized, in the hands of villages and under local control. Given the environment and the nature of the inundation, it could be no other way.

FOR MUCH OF the second millennium B.C.E., Assyria, in what is now northern Iraq, had been a satellite of the more powerful kingdoms of Mitanni to the northwest and Babylon to the south. King Assuruballit (circa 1366–1330 B.C.E.) and his successors reestablished Assyrian independence and founded their capital at Assur, the home of the national shrine. Assur stood above the Tigris but was far from the major centers of population and was short of good agricultural land. A later monarch, Shalmaneser I (1273–1244 B.C.E.), founded a small town at Kalhu, modern Nimrud, near the confluence of the Tigris and Great Zab rivers. Assurnasirpal II (883–859 B.C.E.) chose this obscure settlement as the site for his royal palace, with an outer wall encompassing about 890 acres (360 hectares). An inscribed stela describes how the proud monarch threw a great feast to celebrate the completion of his residence. He claimed to have entertained sixty-nine thousand guests, among them five thousand distinguished visitors, fifteen hundred officials from throughout his domains, and forty-seven thousand people brought to Kalhu to work on-site, as well as the local inhabitants. Many of the imported laborers may have stayed on to live and work in the new city, which would have created an immediate food shortage.[5]

Modern estimates suggest that, without irrigation, the land around Kalhu would have supported about six thousand people. Assurnasirpal cast his eyes on the low-lying alluvial soils near the Tigris and commissioned large-scale irrigation works. He boasted grandiloquently, "I dug a canal from the Upper Zab, cutting through the mountain to its summit, and called its name *Potti-Hegalli*. The meadow-land by the Tigris I irrigated abundantly and planted gardens in its area. All kinds of fruits and vines I planted and the best of them I offered to Assur my lord and to the temples of my land."[6] Traces of this imposing canal can still be seen in a rock-cut channel along the Great Zab. A tunnel, constructed by sinking two vertical shafts and joining them with horizontal borings, passes through a prominent bluff. The canal itself may have originated as far as 10.5 miles (17 kilometers) upstream, the tunnel serving as a regulator that irrigated about 9.5 square miles (17 square kilometers). Assurnasirpal tells us that he devoted the land to fruit and vines, but these were luxuries. Much of the irrigated acreage must have grown barley, perhaps feeding as many as twenty-five thousand people in Kalhu.

Assyrian monarchs obsessed over conquest, control of subject territories, and tribute gathering. Their policies depended on efficient

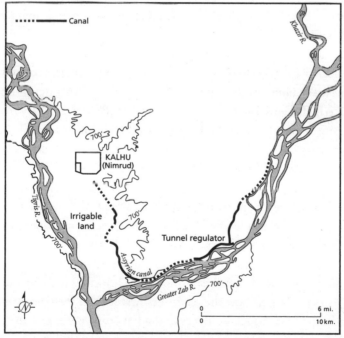

Figure 8.2 *Assyrian irrigation at Nimrud. (After David Oates,* Studies in the Ancient History of Northern Iraq, *p. 46. Courtesy: British Academy)*

administration backed by force and on the deployment of thousands of people, food supplies, and weapons when and where they were needed. If efficient government required the building of irrigation works on a large scale, then the workers to do it, usually prisoners of war and slaves, were ready to carry out the bidding of skilled water engineers, who eyed the terrain and set the line of canals and feeder channels. In many respects, Assyrian irrigation works on a grand scale were a form of conquest by forced labor, which melded well with the mind-set of ambitious kings.

The royal capital moved from Kalhu to Sargon II's short-lived city at Dur Sharrukin (Khorsabad), upstream of modern-day Mosul, during the eighth century. Sargon's son, the boastful Sennacherib (704–681 B.C.E.), promptly moved to Nineveh, whose dusty mounds lie across the Tigris from Mosul. There the capital remained until the Assyrian Empire collapsed and the city was sacked in 612 B.C.E. In its seventh-century heyday, Nineveh covered about 1,800 acres (750 hectares). We have no idea how many people lived there, but Sennacherib undertook massive irrigation works to support them, in part because of his ambitious building programs.

Sennacherib's irrigation works tapped the water resources of the entire area between Nineveh and the nearby mountains. An inscription on a cliff face high above the place where the Gomel River emerges from the mountains boasts of how the king diverted water into a canal that ran cross-country to join the Khosr River, about thirty-one miles (fifty kilometers) northeast of Nineveh. This staggering project involved the building of a stone aqueduct across a wadi near the modern village of Jerwan. "Sennacherib, king of the world, king of Assyria [says] . . . I caused a canal to be cut to the meadows of Nineveh. Over deep-cut ravines I spanned a bridge of white stone blocks. Those waters I caused to pass over upon it."[7] This is the earliest known aqueduct in the world that spans a single wadi, not a series of ravines. Excavations by Thorkild Jacobsen and Seton Lloyd in 1933 showed how the engineers first leveled the deepest part of the wadi, then laid rough boulders and a flat stone pavement to support the columns of the aqueduct. The foundations and columns rose in steps until just below the canal level. A layer of concrete formed the bed for a carefully graded stone pavement with parapets over which the water flowed. Sennacherib completed the canal and aqueduct in about 695 B.C.E. He also built other canals and altered watercourses on a large scale. The king himself said that the Khosr River's water irrigated orchards in the hot season. During the winter, the same waters irrigated one thousand fields for grain crops around and below the city.

None of these stupendous works, nor, indeed, the building of Nineveh and other cities, would have been possible without conquest and the draconian ways in which the Assyrian kings maintained control of areas like the Middle Euphrates Valley and Syria. They deported people from their conquered homelands and deployed them as hostages and laborers. This was the only way in which they could undertake their grandiose building plans, for the Assyrians themselves lacked the manpower to dig long canals and build citadel walls. Thousands of deportees labored for the Assyrian kings and provided the water to irrigate their lands. This may have been one of the ways in which Assyrian engineers learned of the *qanat*.

THE *QANAT* IS an ingenious and highly effective way of mining groundwater by using gravity.[8] The name comes from the Akkadian word *qanu*, meaning "reed." A *qanat* is a sloping tunnel dug into an alluvial fan or a slope with natural water seeps, driven deep into a hillside until it reaches a spring or groundwater underground. The tunnel has to run downhill, so that the water flows smoothly, then emerges at ground level. The user can then divert the water into canals for domestic use or for irrigating fruit trees or fields for cereal crops.

A *qanat* (also called a *kariz*) is not easy to build. A Khursani proverb from Iran remarks that "snake charmers, lion tamers and kariz diggers very seldom die in their beds."[9] *Qanat* building developed into a specialized craft, carried out in historic times by *mughanis*, who passed their knowledge from father to son. First, the *mughani* sank a trial vertical shaft that established the presence and depth of the water table. Exactly where to dig was a matter of observing the topography, the gradient, and local vegetation that offered clues as to what lay below the surface. The builder also took into account the location of the land to be irrigated or the settlement needing water.

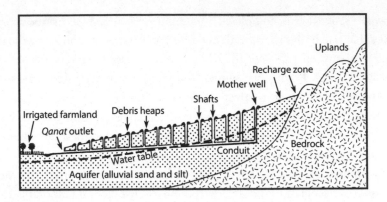

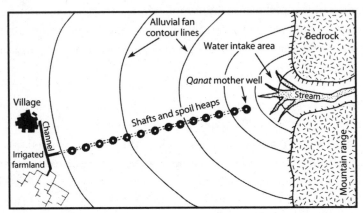

Figure 8.3 *A cross section and plan of a* qanat. *(After Daniel Hillel,* Rivers of Eden, *fig. 9.1. Courtesy: Daniel Hillel and Oxford University Press)*

When the trial shaft reached groundwater, the *mughani* worked out whether the water flow was constant and in an impermeable layer. If all was well, the shaft became the mother well for the *qanat*, the point from which the alignment and gradient of the system was established. The slope angle was critical. Too fast a current eroded the walls of the tunnel, and it collapsed. An unnecessarily shallow gradient caused the water to pool and led to a sluggish delivery. Once he was satisfied with the gradient, the *mughani* then started digging a tunnel, usually about 4 feet (1.3 meters) high and up to 3 feet (around 1 meter) across, starting at the point where the water would emerge on the surface. He sank vertical shafts about 2.5 feet (0.75 meters) in diameter from the tunnel to the surface at irregular intervals. These provided ventilation and allowed the diggers to lift soil to the surface. Digging the tunnel was a relatively straightforward, if hazardous, process, especially if it passed through gravel. Great care was needed to prevent a sudden rush of water when a digger broke through to the base of the mother well. It was not unknown for workers to drown in a sudden flood.

No one knows when this simple (but astoundingly effective) water technology came into use. Almost certainly, *qanats* were an offshoot of both well-digging and mining, perhaps resulting from an accidental encounter with groundwater that flowed out of a mine tunnel. The earliest *qanats* probably date to well before the eighth century B.C.E., perhaps the work of village farmers in the Lake Urumiya region of northwestern Iran. At first, they were little more than a local solution to chronic water shortages in a semiarid environment, until the Assyrian king Sargon II (721–705 B.C.E.) decided to march against the kingdom of Urartu in 714 B.C.E. He wreaked havoc over a broad region around Lake Urumiya, where he admired what appears to have been a *qanat* built by the king of Ursa, a town near the lake: "a main ditch *(palgu)* which carried flowing waters . . . and waters of abundance he caused to flow like the Euphrates. Countless ditches *(atappus)* he led out from its interior [and . . .] he irrigated the fields." Having admired the *qanat*, he promptly destroyed it. "I blocked the outlet of the canal *(hiritu)*, the stream [which was] his reservoir, and turned the fresh waters into mud."[10] Unfortunately, the tablets that record Sargon's campaign are very incomplete, but there can be little doubt that he is describing a *qanat*.

Qanat technology appears to have traveled back to Assyria with Sargon, but may not have been used widely. However, Sargon's son and successor, Sennacherib, tapped the water of the Wadi Bastura, constructing a tunnel through some low hills to convey the waters of three rivers and several springs to the center of the town of Erbil, "the residence of Ishtar, the exalted lady."[11] The tunnel diggers sank shafts at intervals, just like those for *qanats*. Since the Wadi Bastura dries out in summer, it seems entirely plausible that the tunnel also tapped groundwater, for it flowed year-round.

From the air, *qanat* tunnels, with their shafts, look like lines of rings, almost like holes left by stitching. Some have as many as three hundred shafts. *Qanats*' lengths vary greatly. Some are a few hundred yards long; others extend for 5 to 10 miles (8 to 16 kilometers). Tunnels as long as 20 miles (32 kilometers) are not unusual. The largest known *qanat*, at Gonabad in Iran, runs for nearly 22 miles (35 kilometers). The tunnel reaches groundwater 980 feet (300 meters) below the surface, so deep that the shaft has three steps. Most tunnels go no more than a third as deep. The flow varies enormously. One *qanat*, at Sarud, 250 miles (400 kilometers) west of Tehran, produces 20,500 gallons (76,600 liters) of water a day. That is as much water as was produced by the four aqueducts that once supplied Roman Lyon, in France. A small *qanat* can produce as much as 114,000 gallons (442,000 liters) a day, sufficient for a small village. Lesser flows of about 43,200 gallons (163,500 liters) a day are more common, with major fluctuations in the dry season or in drought years. Most larger *qanats* produce about half the discharge of a Roman aqueduct, but they served a different purpose. Unlike aqueducts that service cities and towns, often providing water for civic amenities like public baths, *qanats* supply water needed for subsistence and survival. Thus, founding a village started with the *qanat* digger. Once he had located an underground water source and dug a tunnel, a community would appear around what was, effectively, a new oasis.

Qanats are so effective that they are still big business in Iran. Until 1930, Tehran's civic water supplies came from just twelve *qanats*. In 1961, a French water engineer calculated that about half of Iran's land was watered by *qanats*, with a total of about forty thousand in the country—estimates vary. Twenty-two thousand of them produced nearly 13 million gallons (49,210 million liters) of water daily, the equivalent of a large river. Many are still in use in an era of diesel pumps, so we know quite a lot about them. Unlike pumps, *qanats* are gravity devices, which depend on a relatively constant water table. The height of the water table controls the length of the tunnel and the amount of water it delivers. This contrasts dramatically with the much greater, usually constant water flow drawn from a deep well with a diesel pump. If the groundwater level falls, then the well can be deepened again and again to ensure a reliable flow, whereas a *qanat* is in trouble. But mining groundwater using a *qanat* is a self-sustaining operation, for the tunnel takes less water out of the ground than is replaced naturally. The lesson of history is clear: You cannot mine groundwater to expand human settlement by pumping and reasonably expect the settlement to be permanent. Pumps and wells in combination are not self-sustaining devices.

Complex rules surrounded water supplies derived from *qanats*. Today, individuals own many *qanats*. The rotation of water from a *qanat* to different fields can be on a permanent basis or be agreed upon from one year to the next. Land at a higher altitude has the right to receive water before that at lower elevations. Shares of *qanat* water can be subdivided, then subdivided again, even bought and sold, which creates a fertile environment for sometimes passionate quarrels, even violence. Over the centuries, a body of law has grown up around *qanats*, regulating, for instance, the distance they can be built one from another. Nevertheless, they have remained a staple of water supplies over much of Iran and much further afield for thousands of years.

The *qanat* was an Iranian institution for centuries before Sargon set foot in Urartu. And they remained virtually unknown elsewhere long after the Assyrian Empire had passed into history. Then Cyrus the Great forged the Achaemenid Empire after 550 B.C.E. Cyrus subdued Media, Lydia, and Babylon. His son Cambyses II conquered Egypt in 525, the only remaining independent state of any size in the Near East. Cyrus was an adept administrator as well as a conqueror, as were some of his successors, like Darius I, and it was under Achaemenid rule that *qanats* came into widespread use. They appeared in Syria, where they are still in use, as they are in Oman, where they are known as *aflaj*, and Yemen, where they are called *sahzidj*. In the Holy Land, potsherds show that *qanats* were in use as early as the Persian occupation (537–332 B.C.E.) and were in widespread use during Roman times. They are also found in Saudi Arabia and Pakistan. Achaemenid water-management practices spread into Egypt, where *qanats* were used in the Kharga Oasis, west of the Nile, after Cambyses's time. The device was so effective that it ultimately spread across North Africa, even into Spain, probably in Islamic hands. Spanish conquistadors introduced *qanats* to Mexico after the conquest, where they are no longer in use.

BOTH THE ASSYRIANS and the Achaemenids were conquerors on a grand scale, the culmination of many centuries of usually abortive efforts to forge larger kingdoms out of what were basically tribal societies. In Mesopotamia, the Kassites, a tribe from the Zagros Mountains in Iran, gained control of Babylonia in about 1531 B.C.E. and turned a patchwork of warring city-states into a territorial state. Kassite rulers started ambitious canal and irrigation systems in the south that were expanded by Babylonian kings, who sacked Nineveh and destroyed the Assyrian Empire in 612 B.C.E. They continued opening up huge new areas for settlement, a process that involved parceling out acreage into large estates. The trend toward thinking big about irrigation continued after Cyrus captured Babylon by diverting the waters of the Euphrates in 539 B.C.E. The Achaemenid Empire controlled more than twenty nations at the height of its power, the king's domains ruled by twenty-three provincial governors known as satraps, who enjoyed considerable autonomy. An efficient infrastructure, including an extensive road system, fostered long-distance trade and an empire whose finances came from tariffs, tribute, and expanded agricultural production.

Alexander the Great overthrew the Achaemenid Empire. At his death in 332 B.C.E., the Seleucid Empire rose out of the jostling for power of his generals, at its height a vast domain that extended from the Aegean Sea to Afghanistan and Pakistan. Riven by power plays and assassinations, the Seleucid domains imploded in the face of Rome's challenge, while Mesopotamia came under the sway of the powerful Arsacid, or Parthian, Dynasty, from northeast Iran, which encouraged much-expanded irrigation and settlement in the valley of the Diyala River, a tributary of the Tigris. Like the Seleucids, the Parthians faltered when confronted by Rome, and they were conquered by the Persian Sassanians, from southwestern Iran, in 224 C.E.[12]

Much-larger-scale irrigation took hold in Mesopotamia during these centuries of conquest and counter-conquest, but these efforts paled against those of the Sassanians, whose abandoned canals and gridlike field systems still cover vast tracts of the land between the rivers. The Sassanian rulers created the last pre-Islamic Iranian empire, which flourished for more than four hundred years. Their domains, known as Iranshahr, encompassed a huge area including the southern Caucasus and

Figure 8.4 *Sassanian waterworks in northern Iran. Sadd I Garkaz is a major earthen aqueduct that led water across the Gorgan River into the ditch alongside it. (Courtesy: Tony Wilkinson)*

Armenia, Iran and Iraq, portions of the Persian Gulf and the Arabian Peninsula, and even some parts of southwestern Pakistan.

The Sassanian Empire used draconian policies that had served earlier conquerors like the Assyrians well. The central government invested massively in irrigation systems that dwarfed earlier water-management efforts. They combined the construction of extensive canal systems and weirs with a deliberate policy of resettling deported populations in areas where the authorities saw potential for agricultural production. There, they founded new towns and undertook mind-bogglingly large, hand-dug irrigation works to support them. Near the city of Shushtar, Shapur I, the second Sassanian ruler, built a weir nearly 2,000 feet (600 meters) long across the Karun River. This stupendous project raised the water sufficiently for it to flow down a nearby canal. Numerous huge irrigation projects culminated in the Nahrawan system, built during the sixth century C.E., which carried water from upstream of Samarra through the Diyala Plains to the Tigris, a distance of more than 143 miles (230 kilometers).[13] The canal was between 98 and 164 feet (30 and 50 meters) wide. The upstream portions of it passed through hilly terrain. A combination of Nahrawan and Diyala water irrigated about 3,000 square miles (8,000 square kilometers) of farmland northeast of Baghdad. The irrigation works extended west of the city, where a series of transverse canals linked the Euphrates to the Tigris. These channels delivered too much water, because of sluggish drainage, so the water table rose too high. Acute salinization affected many areas between Baghdad and Babylon as a result. Many thousands of acres were abandoned. The area was almost deserted by 1500 C.E.

One of the best-preserved Sassanian landscapes survives on the Deh Luran Plain, in southwestern Iran. The plain lies near the Iraqi border, about 124 miles (200 kilometers) southeast of the Diyala region. This is a semiarid steppe environment with irregular winter rainfall. Sassanian sites stand out in the region, clustered around newly constructed water-control works in the upper reaches of the Mehmeh and Dawairji rivers and in the nearby piedmont zone. In the river zones, Sassanian water engineers used a modification of the *qanat* system. They ran the tunnels parallel to the deepest river channels, at distances of up to 164 feet (50 meters), collecting water as it percolated through the riverbanks into the *qanat* system. Some of these tunnels ran for a few hundred feet, others for distances of up to 1.3 miles (2 kilometers). The intervening soil acted as a fine screen that kept out river silt and other debris. Ingeniously, the *qanats* collected clean water free of most particles. Fewer minerals reached the fields, too, which reduced the salinization rate.

Until the Sassanians came along, the arid piedmont was largely uninhabited. Instead of importing water from outside, their engineers used low terrace walls of undressed stone on hillsides, spacing them closer in steeper terrain. These walls not only retained soil but also conserved runoff from higher elevations, which made for efficient seasonal cultivation of terraced fields. Check dams built across drainages that ran across the piedmont retained soils and stored runoff that soaked the relatively deep soils behind them. The terraces and check dams clustered in distinct units alongside small dwellings and what appear to be cattle pens. Calculations based on the extent of each field system hint at about 10 to 25 people living in each one, whereas between 80 and 240 individuals may have lived in hamlets practicing irrigation agriculture near the rivers.

Another ingenious device that worked well on the piedmont was a system of canals fed by a series of springs on high ground. The springwater was diverted into canals, each one carefully following the land's natural contours to maintain gentle, efficient grades. Small aqueducts carried water over deep, secondary drainages. Elsewhere, in the northern parts of the system, the engineers built towerlike masonry structures constructed against, and into, steep drainage banks.[14] They stood (and stand) about 900 feet (some 270 meters) apart. The canalized water fell into the towers and dropped about 21 feet (6.5 meters) into a continuation of the canal system. Below the towers, the canals followed carefully engineered gradients. The water served the domestic needs of the communities downstream and also irrigated land that was otherwise useful only for seasonal dry farming. But what were the towers used for? At first, the excavator, James Neely, thought that they were designed to ensure water flow. Then he excavated one and found that it was a simple water mill for grinding grain, a type now commonly known as the Arubah penstock. Such devices deliver water through a narrow orifice at the bottom of the tower, which drives a milling wheel by means of a jet of water under pressure. These are particularly useful in places where stream flows are limited, as they are over much of the Near East. The power developed is proportional to the head of pressure in the drop tower, which makes efficient use of smaller volumes of water.

Across the Deh Luran Plain, Sassanian villages lay in strings along irrigation canals, with households dwelling in bands on either bank. Neely estimates that about seventy-five thousand people lived on the plain during Sassanian times, or about seventy-five people per 0.25 square miles (1 square kilometer), which is roughly consistent with the known carrying capacity of the land and a testimony to water management of the day.

Sassanian irrigation systems in Mesopotamia itself involved large-scale public works that transformed the landscape. The Assyrians, with their aqueducts and tunnels, had started the process, and it continued under later empires, especially the Sassanians. The Sassanian kings used prisoners of war to labor on the projects. Shapur I defeated the Roman emperor Valerian at the Battle of Edessa in 260 C.E. The emperor died in captivity; many of his captured soldiers labored on irrigation canals.

During a short period in the sixth century C.E., a series of Sassanian kings undertook enormous irrigation works between the Tigris and the Euphrates. Archaeologist Robert Adams has calculated that the realities of low river levels during the fall planting season meant that only about 4,000 to 4,600 square miles (10,400 to 12,000 square kilometers) of land could be brought under cultivation annually using the waters of the Euphrates. The Sassanians brought an area at least five times larger than that under at least sporadic cultivation. Even allowing for fallowing, the cultivated area was at least double that of earlier times. The only way they could do this was by planning a comprehensive irrigation system that tapped the waters of the Tigris on a large scale, as well as earlier Euphrates sources. Ambitious works were expensive but imperative for a state whose revenues came mainly from land taxes. Usage of the Tigris was never continuous, for the river is dangerous during the high-flood months. Fall and early winter were the times when Tigris water could be more readily diverted. As the flood rose, farmers along its banks made little use of it except to water their fields on the back slopes of the levees.

Harnessing the waters of the Tigris on a large scale was a high-risk venture, made even harder by the swift and variable current. The Sassanians never dammed the river, but their comprehensively planned diversions involved huge canal investments like the Nahrawan canal system in the lower Diyala region, already mentioned, which carried water long distances before irrigating arid flatlands. The grids of canals and fields covered areas larger than anything that could be administered by village elders or even a small city-state. In earlier times, village elders had supervised breach repairs and decided how to allocate water. Now the sheer scale of irrigation works meant that farmers living some distance from water sources would be in big trouble if multiple failures of weirs or other works occurred further up the system and far beyond village lands. All of this suggests that some form of highly organized, centralized authority supervised every corner of each irrigation system, requiring substantial resources from the imperial treasury to pay for maintenance and repairs.

This was water management on an unprecedented scale, a system that was geared toward producing maximal fiscal returns from the land. These huge irrigation systems yielded great bounty, but they had fatal weaknesses, not necessarily of their own making. The Sassanian kings lived in a volatile world, surrounded by hostile neighbors and menaced by endless internal revolts. The monarch had no option but to think in the short term; his immediate strength depended on maximum returns and potential booty from sudden conquests. There was a constant tussle for power between the monarch and the landowning nobility. The side effects percolated down to their subjects. Each new demand, each ambitious irrigation scheme, reduced the self-sufficiency of village farmers and those who labored in the fields. Each major canal, each expanding lattice of branch channels, transformed the landscape by enclosing lands behind levees in poorly drained basins and helped obliterate an already sluggish natural drainage. The authorities delivered more irrigation water, but they neglected the all-important drainage so carefully fostered by earlier farmers. The water table rose, higher salt levels resulted, and crop yields declined. As population densities climbed and fiscal demands increased, more land had to be brought under cultivation, declining returns or not. Declines in food production would have been catastrophic.

There had always been safety nets for Mesopotamian farmers—their herds, the rich game, plant, and aquatic foods of the swamps. The draconian expansion of irrigation works in the name of expanded crop yields not only shortened all-important fallow periods but also reduced the amount of land available for grazing livestock. Stubble (the remnants of the crop left in the ground after harvesting) was only a partial, seasonal solution, so the loss of such places as the margins of seasonally waterlogged natural depressions meant that herd sizes shrank just at a time when the population of livestock owners was increasing. For many villages away from cities and towns, declining herd sizes may have meant the difference between survival and death, especially when there was not enough grain surplus to feed livestock.

The huge gains in Sassanian water management faltered in the absence of long-term political stability. The empire's ambitious schemes achieved impressive gains at the beginning in greatly increased revenues, more land brought under cultivation, and much larger food surpluses. But these short-term advances came at the expense of increasing ecological fragility. Productivity dropped sharply, especially in marginal areas; the system lost its flexibility amid periodic crises; and food reserves plum-meted. With these troubles came epidemic disease caused in part by hunger, as well as by catastrophic plague outbreaks between the sixth and eighth centuries C.E. The Sassanian Empire dissolved in the face of an expanding Islam between 632 and 651. A precipitous fall in land use over the Euphrates floodplain and elsewhere began in the ninth century. By the eleventh, as much as thirty-eight hundred square miles (ten thou-sand square kilometers) of what Robert Adams has aptly called the "Heartland of Cities," the land between the rivers, the hearth of early civilization, was an abandoned, salt-infested no-man's-land.

CHAPTER 9

The Waters of Zeus

And there by the last row are beds of greens,
bordered and plotted, greens of every kind,
glistening fresh, year in, year out. And last
there are two springs, one rippling in channels
over the whole orchard—the other, flanking it,
rushes under the palace gates
to bubble up in front of the lofty roofs
where the city people come and draw their water.

                                                     Such

the gifts, the glories showered down by the gods
On King Alcinous' realm.[1]

WATER—THE GIFT OF THE GODS: The message resonates down the centuries from the stanzas of Homer's *Odyssey*. Soft west winds brought moisture to a parched land and to Phaeacia, the mythic kingdom of Alcinous. The travel-worn Odysseus gazed with wonder on the palace orchard, where fruit always ripened, where droughts never sucked moisture from the soil. I remember reading of the garden when sailing close to the Peloponnese, perhaps close to the legendary Phaeacia, on a hot summer's afternoon. Steep rows of abandoned terraces climbed the parched hillsides, abandoned by farmers now dependent on pumps and boreholes. The villages where many of them still lived huddled close together, the white or honey-colored houses separated by narrow alleyways wide enough for a donkey to pass. There was none of the lush richness of Alcinous's garden here.

Weeks later, on a cold September evening, we walked a short distance inland from our anchorage on the island of Euboea, northeast of Athens. We climbed a low ridge and came across the village taverna, where a bright hearth glowed. Silence fell as we walked in on a group of weather-beaten farmers drinking ouzo, men with their harsh lives etched on their faces and into their leathery hands. Soon we were talking away in a mixture of English, rudimentary Greek, and sign language amid gales of laughter that dissolved the chasm between our worlds. Here, again, I was reminded of the Homeric garden and the unending struggle between farmers from another era and the land.

This struggle to survive here goes back millennia. From the beginning, Greek farmers had to rely on highly variable, seasonal rainfall, just like villagers elsewhere in the semiarid eastern Mediterranean world. In small villages away from the great rivers of Egypt and Mesopotamia, water management was as old as farming itself, a matter of balancing seasonal, and often irregular, rainfall with the need to water animals, crops, and humans over long dry seasons. All of these relationships between humans and water functioned over thousands of years, largely independent of state control. Huge areas of the Near East depended entirely on water from springs and unpredictable rainfall, on rare perennial and, more often, seasonal watercourses, and on aquifers, mined from an early date with wells, tunnels, and, later, *qanats*. Some of these water-mining technologies go back to the earliest days of farming. A recently discovered well on Cyprus dates to between 7000 and 9500 B.C.E., the earliest known such structure in the world. No question, an expertise with farming in arid landscapes passed westward to Greece with the very first cultivators well before 6,000 B.C.E. No cities, no great leaders, were involved. Compared with, say, the Sassanians' tightly controlled and centralized irrigation systems, village water management was much more stable and had a much longer life span. Cooperation and self-regulation worked in small-scale societies where competition wasn't an issue. Self-sustaining methods were the inconspicuous hydrological backbone of ancient Greece.

The Greeks faced many of the same challenges as people in drier lands to the east, with the difference that the winter rains could descend with great violence, often for days. For months on end, they had to manage their water supplies with great care, but when great storms came, they couldn't just enjoy the rain passively. They had to develop ways of draining away the excess.

Good storm drainage was a necessity, but water conservation was of paramount importance all year. Two sources provided a degree of water security. One could dig a well or construct a cistern, the latter often as part of one's house. I remember being mesmerized by the still-intact house cisterns in the heart of the Byzantine fortress on the summit of Monemvasia, in the southern Peloponnese. There was no other water supply than rainfall for this heavily fortified castle. Cisterns worked surprisingly well. According to one estimate, the inhabitants of one house in the Greek and Roman city of Morgantina, in Sicily, would have had minimal supplies for a year if they had captured but half of the rainfall that fell on their roof.[2] Cistern technology was a foundation of Mediterranean life, and sometimes on a large scale. The rock-cut cisterns high above the Nabatean city of Petra, in what is now Jordan, harvested runoff from nearby slopes for centuries. No one knows how the cistern was invented. Classical historian Dora Crouch speculates convincingly that they may have originated when someone dug a hole for mud brick, then hit on the idea of lining it to make a permanent water-storage place. From there, it would have been an easy step to a roof and a simple collecting system.[3]

Like many Mediterranean peoples, the Greeks had an intimate knowledge of their environment, of the weather, of grass cover for grazing, and of likely places to find water. They were expert geologists, although, of course, they did not think of themselves as such. Their knowledge was that of experience. The fig, the rosemary bush, and mosses offered clues as to subsurface water. Occasionally, they called on water diviners to work their magic. The farmers knew that permeable limestone layered with dense clay would yield water collecting along the seam between the two, which would appear as a spring or seep on the surface. Throughout Greece, the development of water resources, and, indeed, the ownership of them, came not through divine right but through increasingly sophisticated ways of engineering waterworks, and also through conscious decisions about the legal issues of ownership, maintenance, and control. Here, water came to the settlement rather than the settlement to the water, especially with the systematic exploitation of underground water in limestone formations and with the use of pipelines as well as open channels. During the second millennium B.C.E., the Minoans of Crete pioneered the use of aqueducts in rugged terrain.

Arthur John Evans, the discoverer of Minoan civilization, never set foot on Crete until after his fortieth birthday. The son of a wealthy papermaker, himself a respected archaeologist, the restless Evans spent much of his early life in the Balkans, where he walked high mountains, wandered through deep forests, and made what he called "most weird journeys beside people in sheepskin mantles and hats like Cossacks."[4] An incurable romantic, he first arrived on Crete in 1894 in hot pursuit of the forgotten civilization that had preceded the Mycenaeans, recently unearthed on Greece's Plain of Argos by Heinrich Schliemann of Troy fame. Schliemann had always wanted to dig at Knossos, an olive-covered hillside near the town of Iráklion, but Evans managed to buy the site in 1896, just as a brutal rebellion against the Turkish masters of the island broke out. After four hectic years distributing relief supplies, Evans be-

Figure 9.1 *Part of the Palace of Minos at Knossos, Crete, as reconstructed, somewhat imaginatively, by Sir Arthur Evans in the 1920s. An aqueduct and stone-lined channels provided water to the palace. (Luke1138/iStockphoto)*

gan excavations at Knossos in 1900, and he continued working there intermittently for thirty years.

Evans had no excavation experience whatsoever, but within a few weeks he uncovered a labyrinthine palace with numerous chambers, storerooms, and courtyards—and an elaborate water-conveyance system. He named his newly discovered society the Minoan civilization after the legendary King Minos of Crete. The anonymous architects and engineers of the palace had lain an elaborate water system based on local springs as one of their first acts. A series of conduits and pipelines delivered water to Knossos from the Mavrokolybos spring, about a third of a mile (half a kilometer) southward. A network of terra-cotta pipelines passed under the palace, carefully graded to take advantage of slight slopes. Some of the pipes were nearly ten feet (three meters) underground. Remarked Evans, "Minoan engineers, at the close of the Third Millennium B.C.E., made a practical application of the fact that water finds its own level."[5] Here, as everywhere else, gradient moved water, but with an unexpected refinement. The terra-cotta pipes had collars and stop ridges (ridged circles to ensure a proper overlap between pipe segments). Unlike the parallel-sided pipe sections found elsewhere in Greece, each segment tapered, thereby increasing the velocity of the water and preventing sediment buildup. Knossos also relied on sturdily constructed wells.

Drainage was as important as delivery. The overflow waters of the Central Court and its borders flowed down a large drain. The domestic quarters had a main conduit for sewage surrounding them, with waste flowing in opposite directions from the high point at the southeast corner of a large colonnaded hall. Airholes ventilated the stone channels; manholes provided access for cleaning. Evans remarked, "Their use in connexion with latrines would hardly have been tolerable unless they were pretty constantly flushed. Between September and April the recurring, and sometimes torrential rainfall would have produced this result."[6] He waxed lyrical over a latrine on the ground floor paved with a gypsum floor, where the user sat on a wooden seat. A sloped hole and a duct led to the main drain. A projection partially masked the aperture, perhaps for a vessel used to flush the basin. As Evans himself remarked, Minoan sanitation was of a standard rarely attained anywhere in his own day, an interesting reflection of public hygiene a century ago.

The Minoan achievement is the more extraordinary when one realizes that Knossos only receives between about twelve and sixteen inches (thirty and forty centimeters) of rain a year. At another palace, Zakros, they tapped groundwater from a major spring and conveyed it to a square underground fountain built of limestone and accessible by fourteen steps. Crete's mountainous terrain made aqueducts a logical way to deliver water from springs at higher altitudes. Minoan experts combined closed pipes with contour-hugging channels, some extending over distances as long as about 1.5 miles (2.5 kilometers).[7]

The Minoans' hydrological expertise did not vanish with their civilization, for the transportation of water became closely tied to political power. In the seventh century B.C.E., the tyrant Theagenes of Megara, in Attica, kept waterworks firmly in his own hands. He built an aqueduct and a fountain house, sacred to the Sithnidian nymphs, one of whom was said to be the mother of the city's founder. The most remarkable achievement was that of an individual engineer, Eupalinos, son of Naustrophus, also of Megara. In 530 B.C.E., the tyrant Polycrates commissioned him to build a 3,432-foot- (1,030-meter-) long tunnel to bring water from a spring through a 738-foot (225-meter) hill to his town on the island of Samos. Eupalinos appears to have walked over the mountain and set up poles along a straight line, which his workers followed. He dug the absolutely level tunnel from both ends, making a small deviation near the center, perhaps to avoid a natural fracture in the rock. When the two-person teams heard the sounds of each other's tools, they were able to break through at the same spot. A horizontal tunnel was easier to dig, given the simple tools of the day, but it could not operate as an aqueduct. So Eupalinos excavated a sloping channel to one side of the floor. When it became too deep, he dug the deeper portions as a tunnel, using twenty-eight vertical shafts to give the workers access. The water flowed through the channel in clay pipes. Eupalinos could have taken the easy way out and dug channels around the hill. Instead, he persuaded Polycrates to pay for a much more expensive alternative. Perhaps the two men wanted to create a monument to their hydrological vision. They certainly succeeded. The ubiquitous Herodotus marveled at the tunnel and its water channel, "which is twenty cubits deep and three wide, and which carries water from a great spring through pipes to the

Figure 9.2 *The tunnel of Eupalinos, on Samos. The water ran through the trench on the right.*

town."[8] But the most important discovery of all was that of the unique properties of karst.

IT WAS ALL very well digging tunnels through hills, but this was an expensive way of acquiring water, only possible with the resources at the disposal of a despotic ruler such as Polycrates. Fortunately for the Greeks, they lived in landscapes where porous and water-abundant karst abounded. During the first century B.C.E., a Roman engineer, Marcus Vitruvius Pollio (around 80–15 B.C.E.), wrote a ten-volume treatise on architecture in which he summarized the now-lost works of Greek engineers of the sixth to fourth centuries. He devoted his eighth volume to water and wrote of how rain and snow accumulated on Greek mountainsides: "Afterwards in melting, it filters through fissures in the ground and thus reaches the very foot of the mountains from which gushing springs come belching out."[9] He put his finger on the basis of much Greek water management: the unique qualities of karst formations. (Karst is terrain of porous limestone containing deep fissures and sinkholes, marked by underground caves and streams.) Vitruvius had ample grounds for respecting Greek water engineers. Between the eighth and fifth centuries B.C.E., the Greeks were building cities that were increasingly complex and that required careful management of water sources, distribution, usage, and discard.[10] That they were able to do so for centuries is a testament to the remarkable skill of their water experts and especially to their geological expertise.

Carbonate rocks define much of the Mediterranean landscape, laid down and uplifted over millions of years when the sea alternated with dry land. These strata consist of limestone layers, dolomite, or marble. Karst forms where limestone interacts with water to create sinks, ravines, and other surface features, as well as underground water tunnels. It most commonly occurs where soluble rock lies under a permeable but insoluble formation like sandstone. Water flows underground in aquifers through karst, some of it accumulating in conduits such as cave systems, some of it more diffuse, and much of it seeping through joins in the rock. An irregular and usually discontinuous water table can form multiple layers, while springs are common on the surface. Much of the water used by ancient Greek villages and cities had a high calcium carbonate load in the form of sinter, fine dust resulting from the corrosion of limestone. Sinter accumulation was a constant problem in their pipelines and posed a serious maintenance challenge.

The shafts and channels in karst formations act as natural pipelines; dolines (closed depressions) and sinkholes form places where surface water drains into the rock through fissures. Caves, many of them deep underground, often contain springs and also sometimes serve as natural pipelines. All of this makes a superb reservoir for drinking water. Karst formations abound in springs, which usually appear at the base of a mountain, where a layer of soluble limestone abuts an impermeable stratum of stone or clay. Dora Crouch and others have referred to Greek mountains as giant water towers, which provided high infiltration and low runoff, with the karst even forming water tables at high altitudes— the equivalent of humanly constructed Maya "water mountains" in Central America.

Karst yields perennial rivers and springs, both of which can be tapped to supply water over considerable distances. Such springs are not uncommon. In the southwestern Peloponnese alone, there are two major and 118 minor springs, as well as twenty-six rivers that lend themselves to simple irrigation schemes. The settlement pattern of ancient Greece reflects the widespread distribution of karst water supplies of all kinds, exploited with simple technologies that enabled people to live in all manner of local environments and helped turn Athens from a village into a city.

Attica, with its capital Athens, was home to between two hundred thousand and three hundred thousand people during the fifth century B.C.E. (Estimates vary widely.) The limestone Acropolis, with its shrines and temples, was the focus of the city itself, a place where Mycenaeans used a spring as early as the thirteenth century B.C.E. The Acropolis might have had defensive appeal, but its real attraction was its springs and water seeps. As early as the sixth century B.C.E., large roofed cisterns fed by rainfall formed a square inside the Acropolis wall. These

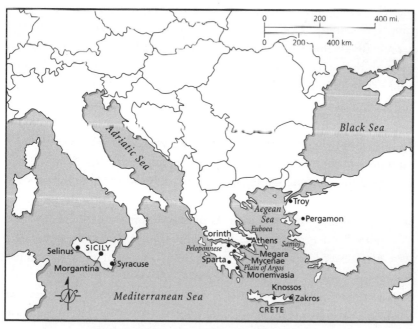

Figure 9.3 *Map showing locations in chapter 9.*

complexes, with their supply and drainage channels, were apparently dug for defensive reasons. A series of springs and some water-bearing caves girdle the Acropolis, formed by percolating water reaching a layer of relatively impermeable schist and marl, then appearing on the surface in the form of small springs. The Acropolis is, in effect, a huge reservoir. Five additional cisterns dug just north of the Parthenon may have served the needs of visitors during major festivals.

The natural springs of the Acropolis were insufficient to satisfy the water needs of the rapidly growing city. From the earliest times, the Athenians clustered at its foot dug shallow wells or pits to tap the groundwater. They used fired clay pipes to draw water from the springs and seeps on the north side. A network of wells developed throughout Athens, reaching the outskirts of the city.

These supplies, too, proved inadequate for the growing city. In about 510 B.C.E., the tyrant Peisistratus and his descendants built an aqueduct that carried water from the Kephissos spring, at the foot of Mount Pentelicus. A second line from Mount Hmettus joined the first in a large reservoir. The aqueduct spanned a distance of over 4.5 miles (7.5 kilometers), ending in the city center near the Acropolis; much of it was a tunnel, reaching a depth of 46 feet (14 meters) in places, the remainder a rock-cut or masonry-block channel. A ceramic pipe carried water through the aqueduct, each section having elliptical covers for maintenance. The ends interlocked tightly with lead-sealed seams. Much later, the second-century B.C.E. Roman occupiers of the emperor Hadrian's day commissioned another aqueduct, which ran from a nearby mountain to a reservoir on Lycabettus Hill. This was effectively a _qanat_, with air shafts at 98- to 131-foot (30- to 40-meter) intervals. Some of the tunnels were large enough to stand in, but in most of them the workers crawled on their hands and knees. Both aqueducts were enormous construction projects by the standards of the day and must have involved slave labor. Underground pipes carried the water through the city. There were also sewer lines and large storm drains, some of them still in use today.

Water was so important to the Athenians that water management became part of the legal system. The statesman Solon became archon (magistrate) in 594 B.C.E. and reformed both politics and the economy. He promulgated laws that stood the test of time. The Greek-born Roman author Plutarch (46–120 C.E.) described how Solon "made a law, that, where there was a public well within a hippicon, that is four stadia [2,300 feet/710 meters], all should use that; but when it was further off, they should try and procure water of their own; and if they had dug 10 fathoms [60 feet/18.3 meters] deep and could find no water, they had liberty to fetch a [pitcher] of six choae [5.3 gallons/2 liters] twice a day from their neighbors; for he felt it prudent to make provision against need, but not to supply laziness."[11]

LOCAL GEOLOGY NOURISHED other Greek cities. Corinth and its successful colony at Syracuse, on the coast of eastern Sicily, also achieved great power and prosperity in part through their clever use of abundant water supplies.[12] Corinth began at the foot of a hill called the Acrocorinth, where springs abounded on the north slope. Both here and at Syracuse, layers of porous rock covered impervious clay strata, where water collected and could be steered into what were effectively underground aqueducts. As early as the sixth century B.C.E., Corinthians tunneled back into the clay under the conglomerate to tap the water seams for fountains. Three of them were in the area of the agora (marketplace). The Cyclopean and Peirene springs lay at right angles to each other, fountain houses that fed cisterns tucked under the conglomerate. Later, a common tunnel fed both fountains from long channels that tapped the karst of the Acrocorinth. The most important spring was the Peirene, a natural outflow point that probably looked like a cave. At first, the water was near the surface, but later generations extended and deepened the supply channels toward the Acrocorinth. Some tunnels ran for nearly two miles (3.2 kilometers) into the rock. Corinth prospered off its abundant water supplies from the karst, until the Romans under Lucius Mummius destroyed the city after a siege in about 146 B.C.E. He killed all the male inhabitants and sold the women and children into slavery. Julius Caesar reestablished the city in 44 B.C.E., settling it with free Romans, who completely remodeled the Peirene spring.

Figure 9.4 *Corinth. The karst formations of the Acrocorinth in the background were the source of much of Corinth's water. (Recreational-Mike/iStockphoto)*

Syracuse also lay on karst and strata, which provided an abundance of water, the water cutting its own channels through the rock, so much so that the city's water engineers had to manage water rather than find or import it. The first settlement was on the island of Ortygia, where a spring with wild papyrus plants bubbled by the sea, named after the nymph Arethusa, who is said to have dived into the Mediterranean to escape the amorous advances of the god of the Nile. As Syracuse expanded to the mainland, people turned to water supplies in natural grottoes above the fifth-century B.C.E. theater built by Athenian prisoners after the Peloponnesian War.

As the population increased, the engineers resupplied these natural outlets with water from further uphill and from the west. They built branching tunnels and stone aqueducts, one of which, the Galermi, ran for about 15.5 miles (about 25 kilometers) from Mount Crimiti into the city. The aqueduct branched many times as it neared the city, filling a reservoir on the outskirts. The Galermi still flows, but contributes little

Figure 9.5 *The amphitheater at Syracuse, Sicily. Much of the city's water supply came from springs in caves behind the theater. (Cassianus 12/iStockphoto)*

to the present city's water supply, though interestingly, a large-diameter modern pipeline follows much of the course of the old aqueduct. The Syracusans invested heavily in water infrastructure for their prosperous city. They combined pipelines and springs with wells and cisterns for domestic use, using major water lines to maintain at least five major public bathing facilities and numerous fountains.

Both Corinth and Syracuse used a logical strategy of relying on diverse sources, a wise precaution in a Mediterranean climate with seasonal and variable rainfall. At the same time, the engineers analyzed every major investment carefully. The greatest demands for water came during the hot, rainless summer. Each city, indeed all Greek metropolises, had to manage its supplies carefully to last through the season. With no fossil fuels, they also had to rely entirely on animal and human energy to distribute and conserve water. As cities grew, their engineers distributed water from more distant sources to existing communities to amplify the supplies from cisterns, springs, and wells. They were well aware that dependence on a single source involved unacceptable risk.

Diversity of supply went hand in hand with efficient water use. The major cities completed what we would call the ecological loop by draining storm water and wastewater away from the center and using it to fertilize and irrigate crops and fruit trees. In this way, they not only produced more food but also replenished the water table so that springs and wells continued to flow for future generations. This is how some Greek cities lasted for nearly five centuries in harsh terrain with uncertain rainfall. Generations of water experts were well aware of the constraints, stored water as close as possible to its points of use, covered channels to reduce evaporation, and diverted the excess from perennially running fountains for other purposes. They differentiated between potable water and water used for purposes that did not involve ingesting it, such as washing and laundry. They also made careful use of wastewater for industrial activities such as fulling and potting. We have much to learn from their long experience.

Drinkable water accounts for 6 percent or less of modern city usage, so—theoretically, at any rate—it is possible to maintain alternative delivery systems of non-potable water for such purposes as bathing, cleaning, and irrigation. In the United States, a few modern suburbs in Florida have installed dual pipelines, with one supplying drinking water inside the house alone. But such seemingly revolutionary ideas are rarely applied, at best. The inhabitants of ancient Greek cities obtained drinking water from springs and public fountains, as well as from private cisterns. Drains and pipelines carried recycled water outside the city for irrigation and other purposes. The wells in the Athenian agora went as deep as thirty-two feet (ten meters) to acquire drinking water, so most people carried freshwater into their houses from them. Cisterns in houses provided water for most domestic and other tasks. Since cisterns were cleaned annually, the water in them was basically clean and suitable for drinking and cooking if need be, but most families preferred the fresh taste of springwater. Bathing, cooking, and laundry water served for flushing latrines and other domestic purposes. Since no one employed soap or detergents, used water could be recycled with much greater ease.

Most urban families combined water from cisterns with that collected in clay vessels from public fountains to survive comfortably. Greek cities spent the resources to provide water supplies conveniently, so that people could make use of them with minimum effort. Their drainage alone added to the quality of life, with drains that led from each house into the public sewers, which were carefully routed to the nearest exit from the city, so that liquid sewage never collected near houses or in the drains. In the city of Selinus, on the south coast of Sicily, domestic bathrooms included a space, separated by a half wall, with a drain leading to the sewer in the street. Dora Crouch believes that people urinated here, using dirty bathwater to flush the floor through the drain. They deposited excrement in portable chamber pots collected by slaves, and it ultimately became fertilizer on fields outside the city walls. These efforts are comparable to the sewage systems at Harappa and Mohenjodaro, in the Indus Valley, a thousand years earlier, described in chapter 11. Contrast ancient Greece or the ancient Indus with nineteenth-century London or early twentieth-century Calcutta, where stagnant, liquid sewage caused regular outbreaks of typhoid and other diseases.

THE GREEK WORLD changed profoundly after the fourth-century-B.C.E. conquests of Alexander the Great. Before his time, the Greeks' realms encompassed mainland Greece, the Aegean Islands and mainland coasts, and Cyprus. Almost overnight, Alexander added enormous tracts of Asia Minor and the Near East, as far east as the Indus Valley. For half a century after his death in 323 B.C.E., his generals fought over and partitioned his empire, before a degree of stability descended over his former domains. The Hellenistic world developed into some powerful states, one presided over by the Ptolemies, the final dynasty of pharaohs in Egypt; another ruled by the Seleucids, briefly mentioned in chapter 8; and Pergamon, in western Turkey. After 200 B.C.E., the growing power of Rome intruded into what was a complex political world, first as a mediator, then as a conqueror, culminating in the conquest of Egypt in 30 B.C.E.

Greeks were never in the majority, but they ruled exceptionally diverse kingdoms and acquired extraordinary wealth. Theirs was now a cosmopolitan world where ideas and technical knowledge flowed freely over wide areas. Powerful rulers like the Ptolemies fostered literature and scientific experiment. Alexandria became a thriving center of intellectual inquiry and technological innovation. Ptolemy I Soter established a library and museum at Alexandria in about 300 B.C.E. that became the equivalent of a modern research university. Most of the scholars came from else-where in the Hellenistic world, to form what must have been a unique intellectual community. Ctesibius, perhaps the head of the museum, was a technological genius, who is said to have begun life as a barber.

Apart from mathematical and philosophical inquiry, much of the in-tellectual activity in Alexandria surrounded military devices, among them catapults and siege engines. The Ptolemies were also anxious to increase agricultural production beyond the levels possible using the traditional irrigation methods of earlier pharaohs. They cast their eyes on the Fayum Depression, a natural basin covering some 656 square miles (1,700 square kilometers) and containing a lake named Birket Qarun, located in the desert some 37 miles (60 kilometers) west of the Nile. The Fayum, known to the ancient Egyptians as Ta-she, or "Lake Land," was once connected to the river during the annual inundation, which brought both water and nutrient-rich silt into the low-lying terrain. After 2025 B.C.E., the pharaohs turned it into a carefully irrigated granary, using tradi-tional irrigation methods. Now, many centuries later, the Ptolemies en-couraged their scientists to develop water-lifting devices to help them reclaim the Fayum, thereby offsetting the high cost of slave labor. Before Hellenistic inventors came on the scene, the only water-lifting device in widespread use was the *shaduf*, a simple leverlike device developed in Mesopotamia and widely used in Egypt, which would lift water a maxi-mum of about 8 feet (2.5 meters), depending on the length of the arm. *Shadufs* are at best small-scale water movers, even when used in large numbers. Sometime during the fourth century B.C.E., but perhaps ear-lier, the waterwheel came into use, a relatively straightforward invention once one realizes that the most efficient water-lifting devices involve cir-cular motions.[13] In their earliest forms, waterwheels were compartmen-talized, hollow wooden containers constructed around a heavy axle, known as a *tympania* and turned by men treading on the circumference or by an ox. Water entered the carefully waterproofed body through openings at the edge as the wheel turned, pouring into a trough through holes next to the axle when the forward end of the compartment reached the horizontal plane. The Egyptian *saqiya* which appeared during the third century B.C.E., was a horizontal version of the compartmentalized wheel, operated by an ox and a simple worm gear, a screw-threaded gear that meshes with a toothed wheel.

Compartmented wheels could lift quite large volumes of water, but only short distances, so a version with compartments in the rim soon came into wide use, long spokes supporting a narrow, hollow rim partitioned into separate cavities at the spokes. As the wheel turned

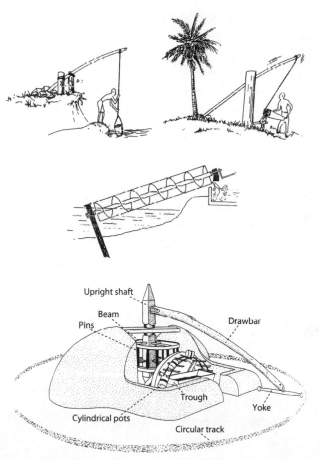

Figure 9.6 *Water-lifting devices. The* shaduf *(top row), the Archimedes screw (center), and the* saqiya *(bottom). (After Daniel Hillel,* Rivers of Eden, *fig. 3.8. Courtesy: Daniel Hillel and Oxford University Press)*

past the vertical, the water in the compartments would pour into a waiting tank.

Waterwheels require animal or human power to drive them. The truly revolutionary development came when an anonymous inventor added paddle wheels on the outside surface and used waterpower to turn the machine. Water-powered lifting devices may have first come into use along the Nile, especially in irrigation canals, where there was a constant flow, rather than in the main river, where the water level changed constantly. Waterwheels delighted many observers with their perpetual motion, like the unknown author in the *Anthologia Latina* who wrote, "It [a compartmented wheel] pours out and scoops up water, discharging on high the stream it carries, and it drinks up a river only to disgorge it. A marvelous achievement!"[14]

Ancient waterwheels of this general type, known as *norias*, are still in use on the Orontes River, near Hama, Syria. They were originally used to feed water into nearby aqueducts and for irrigation purposes. They were widely adopted in Roman times to service public baths, like those described in chapter 10.

Figure 9.7 Norias *on the Orontes River, Hama, Syria. (Johnny Greig/ iStockphoto)*

Archimedes and Ctesibius added to the vocabulary of innovations with logical but revolutionary water-lifting devices. During the third century B.C.E., Archimedes (circa 287–212), the mathematical genius of the ancient world, flourished under the patronage of Hiero II of Syracuse. He corresponded with scholars in Alexandria and visited there. During his visit, he is said to have invented what became known as the Archimedes screw, a device for moving water into irrigation canals, originally developed to pump out the bilges of a large grain freighter. This manually operated device is basically a screw inside a hollow pipe that moves water up a spiral tube as its shaft is turned (figure 9.6). Such pumps could be turned by hand or by someone treading on them. The same screw technology, eventually operated by windmills and later by mechanical pumps, survives today in sewage-treatment plants and fish hatcheries, for the largely self-purging screw is good at transferring liquid with suspended solids. Such devices do not lift water very high, but they are very effective, for there is little friction. The Archimedes screw was such an improvement on earlier technologies that it spread rapidly throughout the Greek world and was widely adopted by the Romans, often used for draining cofferdams and mine shafts.

For his part, Ctesibius developed refined versions of the water clock, for keeping track of the hours, as well as the force pump, which pushed water into cylinders and expelled it under pressure. Force pumps sat in the water and pushed it through tall pipes with the aid of pistons working in two cylinders. The squirting discharge often made a whistling sound. Both pistons and their liners could be manufactured of local materials, which made them easy to build in rural areas. Force pumps could be used for small-scale irrigation. In Roman times, they served mainly for firefighting.

The third century B.C.E. was a time of wide-ranging innovation in pumping and other water technologies, some of which are still in use today with remarkably little change. In Ptolemaic, and later Roman, Egypt, the stakes were certainly high, for these technologies increased agricultural productivity dramatically, turning the Nile Valley into a vast granary for the wider Mediterranean world. For instance, under Ptolemy II and Ptolemy III the Fayum project drained large tracts of marsh by cutting long dikes under the supervision of the engineers Cleon and

Theodorus, between 258 and 237 B.C.E. Cleon was so successful that one of the canals was named after him. As soon as reclaimed land became available, the authorities settled farmers on it to bring the acreage under cultivation. The same villagers carried out maintenance work on the canals during the slack agricultural months between July and November. The scale of these irrigation works, carried out with great speed, is an impressive testimony to the efficiency of the Egyptian government of the day.[15]

The same water-lifting technologies were to spread widely through the Roman Empire, both for prosaic applications such as mining and to provide water to cities for much more ambitious purposes than merely domestic or agricultural uses. All these innovations were, however, in the final analysis, no replacement for gravity, which was capable of delivering water in large quantities without an expensive, continual investment in expendable human labor. And expendable such labor was. Working around water was a task for the lowest of the low, those in bondage or convicts. The astronomer and writer Julius Firmicus Maternus wrote in the fourth century C.E. of work in Rome "that exhausts men with its terrible, unbroken toil. For they are either compelled to draw water from deep wells in their daily task or ordered to clean settling tanks and sewers without rest."[16] Animals freed humans from some of the dreariest work, but in cramped quarters or inaccessible places like ships' bilges or the roofs of bath buildings, humans were the only source. Humans were adaptable, though; people could keep water flowing when height and gravity did not suffice. No Greek, or, for that matter, Roman, engineer ever developed a pump that could rival gravity-driven channels, which is why the Greeks made use of aqueducts and the Romans developed them to even higher pitches of refinement. Greek cities thrived in broken terrain or grew up around high outcrops like the Acropolis. In places like Pergamon, in western Turkey, they used not only aqueducts but also siphons, which carried water across deep valleys.

ALEXANDER THE GREAT founded Pergamon after conquering the Persians. Between 281 and 133 B.C.E., the Attalid Dynasty ruled over the city with generosity and intelligence. The city began as a defensive settlement built on a high acropolis, overlooking a fertile valley. Strategically placed on international trade routes from Asia to Greece, and later Rome, Pergamon became a distinguished city, remarkable as a center of learning and for its library, which rivaled that of Alexandria. Its rulers occupied a hilltop and faced the challenge of bringing water to higher ground. They did this by building five aqueducts of a level of sophistication unmatched in the Hellenistic world.

Only nine Hellenistic aqueducts are known, those at Pergamon being of an unmatched grandeur.[17] The Attalus and Demophon aqueducts were about 12 miles (20 kilometers) long. The Madra Dag aqueduct extended over only 2.3 miles (about 3.5 kilometers) but spanned a valley 659 feet (201 meters) deep. The anonymous designer used a powerful siphon to force water to a high elevation. In 1899, a German archaeologist estimated that the water exerted a pressure of 240 pounds per square

Figure 9.8 *The amphitheater at Pergamon, Turkey, whose water supplies arrived via aqueduct. (Sumbul/iStockphoto)*

inch (18.5 kilograms per square centimeter) on the lead pipes that carried the water through the aqueduct, equivalent to the working pressure of steam in railroad locomotive boilers of his day.

Siphons came into use when an aqueduct engineer confronted a valley too wide or deep for a bridge. Most of what we know about them comes from Roman examples. They used bridges up to valley heights of about 164 feet (50 meters) and preferred them generally, as they were cheaper to build, given the high cost of lead pipes. But when bridges were not feasible, siphons were an ingenious solution. The layout was simple. On the source, or "upstream," side of the valley, a header tank received the water. A row of parallel pipelines carried the water down the side of the valley and across the bottom, often across a low bridge designed to reduce the pressure on the pipes. The force of gravity then shot the water uphill through the same pipes into a receiving tank, where it resumed its journey along the aqueduct. Building siphons required expert judgment and careful design, lest the force of the water shatter the pipeline. The Madra Dag installation at Pergamon was a remarkable engineering feat by any standards.

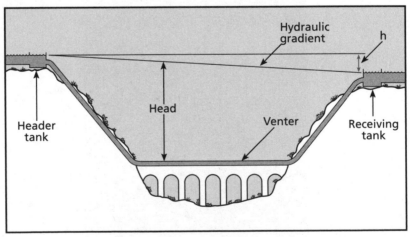

Figure 9.9 *How a siphon works. The force of gravity slingshots water from one side of a valley to the other. (After A. Trevor Hodge,* Roman Aqueducts and Water Supply. *Copyright © 2002 by Duckworth & Co, London)*

The Pergamon aqueducts came into use just as Rome began developing political links with the city, the first stage of what became a close alliance. The Greeks may have refined the idea of the aqueduct from earlier times, but as we will see, it was the Romans who developed the truly long-distance aqueduct, in response to mushrooming urban populations and increased water demand for other-than-basic domestic or agricultural purposes. The Romans may not have been brilliant innovators, but they knew a good idea when they saw one. In the case of aqueducts, they adapted the cumulative experience of millennia to their needs, making them the signature of their enormous hydrological endeavors.

CHAPTER 10

Aquae Romae

GAIUS PLINIUS CAECILIUS SECUNDUS, Pliny the Younger, was a distinguished Roman consul and leader of the first century C.E. He is most remembered for his eyewitness account of the great eruption of Mount Vesuvius in 79 C.E. But we should also remember his country villa, with its bedchambers alive with fountains plashing softly. In one room, "you can fancy you are in a grove as you lie there, only that you do not feel the rain as you do among trees. Here too a fountain rises and immediately loses itself underground."[1] Pliny reveled in the quiet life at his rural villa and the quiet splashing of its artificial streams and fountains. He and his fellow Romans loved water and its perpetual flow, its seemingly endless supply, captured from springs and rivers in cisterns and wells. And, as with everything else, they thought of it on a grand scale, which is why aqueducts large and small are virtually ubiquitous in the lands of what was once the Roman Empire.

The Romans were administrators and soldiers, as well as civil engineers of great ability, but they were not necessarily innovators. Their genius was in observing the practices of others and modifying them for their own purposes, often on a much larger scale. They were experts at surveying and deploying large numbers of workers to tunnel and move earth, to create military camps, to hammer rock and mine far below the surface. Like the Sassanians, they organized water supplies, but with a technology that had advanced little over that of the Greeks and earlier societies throughout the Mediterranean world. For centuries, Rome and other growing communities used cisterns and wells for most of their domestic water supplies, drawing on hydrological philosophies, practices, and traditions that were centuries, if not millennia, old. They learned lessons in water management from the Etruscans and from the Hellenistic world, where, it is thought, they acquired a basic skill in aqueduct construction, the most durable water-management legacy of Roman civilization.

SEXTUS JULIUS FRONTINUS said of Roman aqueducts, "With such an array of indispensable structures carrying so many waters, compare, if you will, the idle Pyramids or the useless, though famous, works of the Greeks!"[2] Frontinus was a gifted public servant of Rome, who served as, among other things, governor of Britain. In 97 C.E., he was appointed water commissioner for Rome, called to a position that had long been a corrupt sinecure. Frontinus was a zealous champion of reform, a Roman of the old school, who wrote technical works as much for his own edification as for the benefit of others. His *De Aquaeductu Urbis Romae* is a treatise on Rome's aqueducts—their names, who built them and when, where they originated, their courses above- and belowground, and where their water ultimately went. He explored the laws and regulations that governed them. This honest and entirely trustworthy administrator knew that aqueducts were enduring symbols of Rome's power and technical prowess. Wherever Rome went, aqueducts went, too. And unlike the Colosseum and other public monuments erected for public entertainment, aqueducts had enduring social value. Over much of Europe and the Mediterranean, they survive as permanent reminders of Rome's imperial might.[3] The aqueduct that still supplies water to Segovia, Spain, was built in 50 C.E., with a spectacular two-story arcade that carries water into the heart of the city. Aqueducts provided water in North Africa, Byzantium, Greece, Gaul, and Germany, to mention only a few locations. The Eifel aqueduct once carried water from the Eifel mountains to the Roman settlement that is now Köln, Germany, a distance of about 59 miles (95 kilometers). The entire aqueduct system, with its outlying spurs, extended over 81 miles (131 kilometers). Aqueducts were Roman architecture writ large.

The aqueduct had a long history before Rome, in the hands of the Assyrians and the Greeks, and in those of the Etruscan city-states of

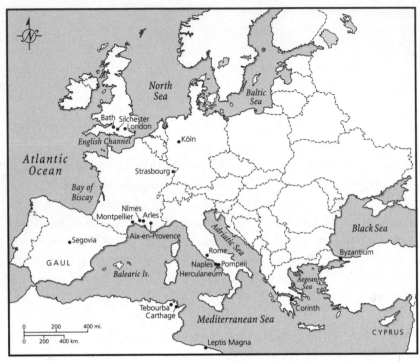

Figure 10.1 *Map showing locations in chapter 10.*

northern Italy. Roman water experts inherited an ancient engineering tradition with roots in many cultures, but it seems likely that many of their early ideas came from the Etruscans, who relied heavily on tunnels known as *cuniculi* (rabbit burrows). The Etruscans tunneled to carry erosive streams underground, to drain marshlands, to irrigate fields, and to water cities. They were adept traders, who may have learned of *qanat* technology from lands as distant as Turkey, for, like *qanat* builders, they dug their tunnels with vertical shafts at irregular intervals.

The first Roman aqueducts brought water to Rome, the earliest being the Aqua Appia, constructed in 312 B.C.E. For centuries, the Romans had relied on traditional water sources: streams, rainfall and cisterns, and the Tiber River. But the city's patricians were a competitive lot, who vied with one another to commission public works. Thus it was that the censor (magistrate) Appius Claudius Caecus supervised the construction of the Aqua Appia, as well as another, much more famous project, the

Appian Way, the first of Rome's great roads that led into the conquered lands of Latium. His engineers surveyed the 12.4-mile (20-kilometer) route of the aqueduct with great care, using Etruscan tunneling methods, with their vertical shafts. The aqueduct channel flowed underground for almost all its length, only emerging on the surface within the city limits, lest enemies try to block it or contaminate the water.[4]

Aqua Appia was just a start. A second aqueduct, Aqua Anio Vetus, completed in 269 B.C.E., also ran underground, but for more than 43 miles (69 kilometers), well over three times the length of its predecessor. Once Rome had secured Italy and defeated Carthage, aqueducts moved aboveground, starting with the third and longest of Rome's water conduits, the Aqua Marcia, built in 144–140 B.C.E. by the praetor (senior magistrate) Quintus Marcius Rex. The funds for its construction came from the conquests of Carthage and Corinth, sufficient to bring water from the Anio Valley, near the modern towns of Arsoli and Agosta, over 52 miles (91 kilometers) from the city. A 5.6-mile- (9-kilometer-) long arcade brought the aqueduct into the city at a height that ensured water supplies even for the highest neighborhoods. When Frontinus measured the city aqueducts in about 97 C.E., the Aqua Marcia supplied about 49,460 gallons (187,600 cubic meters) of water a day, the second-greatest source for the urban area and one of the reasons Rome could become a great imperial city with over one million inhabitants. In the end, Rome boasted eleven aqueducts, which carried some 280,000,000 gallons (1,127,220 cubic meters) a day, all of it transported by gravity. Small wonder that aqueducts became a staple of Roman water distribution.

THE CANADIAN CLASSICIST Trevor Hodge compares Rome's aqueducts to an electrical grid, for they worked on the principle of constant supply, the water never being stored but being used at once.[5] For the most part, the water flowed in an open channel, usually about a third full, although the depth could vary according to the season and the amount of rainfall. As Hodge remarks, Roman aqueducts are best considered as artificial rivers rather than as water mains.

Aqueducts were immensely expensive undertakings that could, and sometimes did, bankrupt a city. They were sources of civic pride, especially

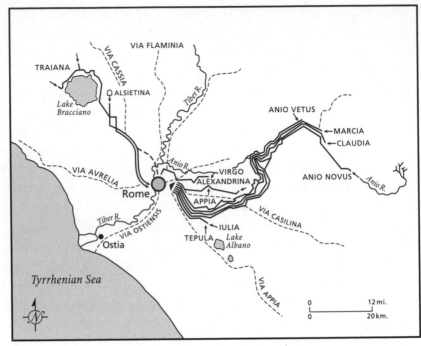

Figure 10.2 *Rome's major aqueducts.*

when the water arrived atop long arcades, like those that crossed the Roman Campagna or the Pont du Gard, which carried springwater on the last stage of its nearly 31-mile (50-kilometer) journey to the city of Nîmes (Roman Nemausus), in southern France. The tri-level bridge, with its fifty-two arches, is 300 yards (275 meters) long and 161 feet (49 meters) high. A road crosses the valley on the middle level; a water conduit 5.9 feet (1.8 meters) high crosses the top one, with a gradient of only 0.4 percent. The aqueduct delivered five million gallons (twenty thousand cubic meters) of water daily, descending only 58 feet (17 meters) over its entire length. Once the water arrived at the storage tank (*castellum*) in Nîmes, it emerged in a public gathering place, where visitors could watch the water flow past. An endless water supply was virtually magical in a world of poverty, hard work, baking sun, and irregular rainfall.

Roman aqueducts typically gathered water through a network of catchment tunnels from several springs. A common reservoir supplied the main channel, the water level maintained at the correct depth by a

Figure 10.3 *Pont du Gard aqueduct, near Nîmes, France. (Photovideo-stock/iStockphoto)*

dam or weir, an adjustable gate controlling the flow through the channel. Such a barrier was important when stream levels rose in winter, or when work on the aqueduct channel was required. Frontinus described the varying quality of the water in Rome's aqueducts, the best coming from the Marcia and Claudia aqueducts. Vitruvius recommended that the engineers "test, inspect, and observe the physique of the people who dwell in the vicinity before beginning to conduct the water, and if their frames are strong, their complexions fresh, legs sound, and eyes clear, the springs deserve complete approval."[6] Newly excavated springs passed muster if their waters left no spots "when sprinkled in a Corinthian vase" or a bronze vessel. Muddy water was to be avoided, sometimes brought down-aqueduct by storms eroding stream banks. Maintaining an even modestly clean water supply required constant vigilance. The builders collected sediment in settling tanks, which were cleaned sporadically, but sinter was a more serious problem. The calcium carbonate incrustation resulting from the sinter's separating from the water had to be chipped laboriously and repeatedly from the sides and bottom of the aqueduct channels.

The preferred method of construction was the covered trench or tunnel, not the bridge, conspicuous as these appeared. They were cheaper to build, avoided the stresses that affected freestanding structures, and were less costly to repair. Most channels were about three feet (one meter) wide and about six feet (two meters) deep. Once they were carved out, the bottom, walls, and cover were built of masonry, unless the channel ran through solid rock. As concrete came into use after the emperor Augustus's reign (27 B.C.E.–14 C.E.) and around the time of Christ, it replaced stone. Whatever the construction, the bottom of the channel received a layer of fine mortar to ease the water flow. The longest aqueduct tunnel in the Roman world was a stretch of the Anio Novus aqueduct outside Rome, which was 1.4 miles (2.3 kilometers) long.

Gravity drove every Roman aqueduct. The path of each one was calculated with sedulous care, to be neither too steep nor too shallow for a steady flow to be maintained. The Roman expertise in aqueduct construction depended not only on engineering skill and the ability of the authorities to marshal large numbers of people to build their waterworks, but also on precise surveys that extended over long distances. Roman engineers were expert surveyors, using a device developed by the Greeks known as a *chorobates*, basically a builder's level with a 19.6-foot (6-meter) beam supported on carefully braced legs. A groove ran down the center of the beam. When water was poured into it and uniformly distributed along its length, the beam was level, allowing the surveyor to project a sight line down it. By using frequent vertical shafts dug into the growing tunnel, the builders could monitor the level from the surface.

The average gradient was about 0.15 to 0.3 percent, but there was considerable variation, with one 3.7-mile (6-kilometer) length of the Carthage aqueduct achieving a gradient of 2.8 percent. The more level the gradient, the slower the flow, while a vigorous flow might erode the edges of the channel. Aqueducts followed the contours of the terrain, passed through hills, and crossed valleys and ravines over bridges or viaducts. As a last resort, the engineers turned to siphons to transport water across deep valleys, as had the Greeks.

Where the Romans excelled was in bringing water across rough terrain or across the wide plains that often surrounded their cities, centered

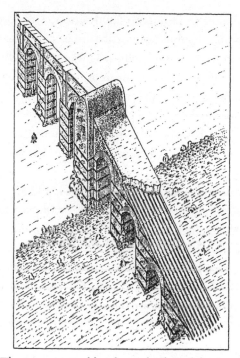

Figure 10.4 *The reconstructed header tank of an elaborate siphon array at Gier Aqueduct, Beaunant, Lyon, France. (After A. Trevor Hodge, Roman Aqueducts and Water Supply Copyright © 2002 by Duckworth & Co, London)*

as they commonly were on an acropolis. They used bridges and arcades to maintain the height of the channel. For a shorter span over a narrow valley or ravine, a bridge was constructed. For much longer runs that could extend for miles, the builders strung together series of arches. These were cut stone or brick-faced concrete, each arch being buttressed by its neighbors. During the first century C.E., they turned to concrete, which they used with untrimmed rocks as the core of the structure, facing it with bricks bonded with mortar or with stone blocks. When a bridge or arcade fell into disrepair, the builders preferred to shore it up rather than replace it, using buttress walls or interior arches that often covered a much more attractive earlier structure.

Arch after arch carried water into the city and into the *castellum*, which served as a distribution point for the open conduits or pipelines that carried water to different neighborhoods, to fountains, and, most important of all, to public baths. Like Greek cities, Roman urban centers depended not on water on demand, as we do today, but on a continuous supply—the Romans expected nothing less. This may seem wasteful in our eyes, but there was no other option. Aqueducts could not be turned on and off. Fortunately, much of the surplus flowed into the sewers and flushed away solid waste, allowing Romans to live at least reasonably comfortably in large, densely populated cities without sewage-treatment plants.

Given the abundance of aqueducts, one could argue that the Romans were enlightened in their desire to deliver potable water to all. In fact, they never built aqueducts purely to supply drinking water or for sanitary reasons. Even cities like Rome had wells and cisterns in most houses. Some cities, like London (Roman Londinium), thrived for centuries without any aqueducts at all. Aqueducts were sometimes commissioned after a city had prospered for decades, even centuries, and were often built largely to supply public baths, a matter of civic pride and imperial magnificence, bathing being a central part of Roman life. Of course, they also supplied water for domestic purposes, as well as for industrial purposes like powering grain mills, for irrigation, and for public toilets. The waterpower they could bring to bear was impressive. At Barbegal, 7.5 miles (12 kilometers) from Arles (Roman Arelate), two aqueducts bringing water to the town became one as they arrived at a steep hill. A sluice enabled the operators to control the water supply to sixteen waterwheels built into the hillside in two rows. Between the late first and late third centuries C.E., the mills ground an estimated 4.5 tons (4.08 tonnes) of flour daily, enough to bake bread for the 12,500 inhabitants of Arles at the time. The greatest consumers of aqueduct water, though, were public baths. Even a modest public-bath complex required enormous quantities of freshwater. In Rome, an estimated 17 percent of the water supply went into public baths, probably a smaller proportion than that of many other smaller cities with large bathing facilities.

EVEN RELATIVELY SHORT aqueducts required long lengths of piping.[7] Pipes had advantages over open channels. They were more tolerant of minor gradient changes, were more flexible to construct, and, if laid in batteries of two or three, could carry a respectable volume of water. Terra-cotta and stone pipes served for at least part of many aqueduct pipelines and were, of course, used extensively by Hellenistic engineers. The Kuttolsheim aqueduct, serving Strasbourg (Roman Argentoratum), used a pair of terra-cotta pipes throughout its 12.4-mile (20-kilometer) length, each 11.8 inches (30 centimeters) in diameter. In Britain and other northern locales, the engineers used timber pipes, often tree trunks with a hole bored through the middle, the joins reinforced with iron collars. In some places, pipes replaced unserviceable sections of a channel, as they did at Caesarea Maritima, in Israel, where a section sank into the soft ground. Pipes solved the problem, but at the expense of a greatly reduced flow.

Above all, however, the Romans preferred lead pipes, not only for siphons and lengths of aqueduct lines but also for urban pipelines and domestic use, so much so that the modern word "plumber" comes from the Latin term, *plumbarii*. (*Plumbum* is the Latin for "lead.") As Vitruvius pointed out, the Romans knew well that lead was poisonous for the "workers in lead, who are of a pallid colour; for in casting it, the fumes from fixing on the different members, and in daily burning them, destroy the vigour of the blood."[8] He strongly recommended drinking water from terra-cotta pipes for this reason. The myth that the Romans suffered from lead poisoning from their water pipes persists, when, in fact, they did not. Their water supplies ran constantly and rarely settled in the pipes; the rapidly forming calcium carbonate deposits in the pipes also formed a barrier between the lead and the water.

Lead had major advantages for pipelines. It was abundant throughout Roman domains, oddly enough most commonly in Britain, where wood pipes were preferred. It was cheap and could be mined or produced as a by-product of silver, although it was very heavy to transport. The soft metal was easily hammered into sheets that were readily bent into pipes, yet it was strong enough to resist serious water pressure. Its low melting point made casting or soldering with lead and tin an easy matter, even if soldering the joints in place at the bottom of a trench must have required skilled work with a hot iron. The pipe sections came in lengths of 10 Roman feet (36 Roman feet equal 35 English feet), created by

Figure 10.5 *A Roman lead pipe from Aquae Sulis, Bath, England.*

melting lead on a flat surface with low sides. By pouring out a set weight of lead, workers achieved sections of more-or-less standard thickness, which were then rolled around a bronze or wooden cylinder to a variety of standard diameters. The edges were then welded, soldered, or hammered, the pipe being installed seam side up to facilitate maintenance. When a junction was required, as they often were in town, the pipe layers simply cut a hole and welded a T-junction, or angled, pipe in place. Enormous quantities of lead serviced Roman plumbing. In the late nineteenth century, archaeologists found a 5,741-foot- (1,750-meter-) long lead pipe, which conveyed water from a reservoir to the Roman Forum. This pipe alone required 257 tons (283 tonnes) of lead.[9]

ONCE THE WATER arrived at the end of the aqueduct, it poured into a large tank where the flow decelerated and, ideally, much of the sediment settled to the bottom. Sometimes there were two or more settling chambers, accessed for cleaning from the side or top. Aqueducts carried large volumes of water, so filtration was rapid at best. Some drinking water from cisterns passed through a circle of amphorae (storage jars) packed with charcoal and sand; grilles and even filtration barriers may have been used to filter aqueduct water on occasion, but they rarely survive.

The aqueduct ended in a *castellum* and setting tanks, which were the most complex part of the supply system.[10] Sometimes the aqueduct first discharged into a large reservoir, usually at least partly dug into the ground and roofed with a concrete vault, especially in places like drought-plagued North Africa, where water storage was essential. Some of these reservoirs were of impressive size. The Carthage aqueduct ended a run of 56 miles (94.3 kilometers) in the huge Bordj Djedid cistern, which was the size of an entire city block, subdivided into eighteen transverse compartments, and had a capacity of as much as 792,537,000 gallons (30,000 cubic meters). This was a major source of supply, but only rep resented a day and a half's discharge from the aqueduct. No aqueduct tank was ever intended to store water for long periods, even those that were major sources of supply. Again, we have to think of the tanks as part of artificial rivers that flowed all the time, year in, year out. Any thought of water storage for peak or off-peak use was totally alien to the Roman mind.

The *castellum* itself was a smaller tank on high ground at the edge of the city, where aqueduct water entered in one stream but left through several pipelines. A few larger users, like public baths, had their own pipelines, but most channels passed to water towers, where their waters then flowed to fountains or individual users. One of the best-known water-distribution systems was in Pompeii, which received an aqueduct during Augustus's reign. The same aqueduct also served Naples (Greek Neopolis: "New City"), with Pompeii on a branch channel that was only some ten inches (twenty-six centimeters) square. The town's distribution system was very elaborate, with the *castellum*, with its oval pool and two mesh screens, lying inside a brick structure. Three exit channels with wooden barriers of different heights carried water out of the pool. The central pipeline of the three served the town fountains; the two others, the baths and theaters and private residences, respectively. The barrier levels show that the town had allocation priorities determined by the water level in the pool. The highest gate was on the private-residence line. Home owners no longer received water when the level fell below their barrier, followed in turn by the baths and theaters. The fountain line was the last to cease flowing, which was logical since it served the largest number of people. Effectively, home owners got no supplies until the other groups received water. Of course, the aqueduct water supplemented that from cisterns and wells, but clearly the public interest was well served.

At Nîmes, the *castellum* was an open, shallow basin 18 feet (5.5 meters) across and 6 feet (1.8 meters) deep. An arc of ten pipelines carried water from the pool into the city, but how exactly the operators controlled the flow and allocation is a mystery. Once outside the pool, the ten pipelines grouped into five pairs, each running between two masonry walls; their ultimate destinations are unknown. The *castellum* appears to have been a popular gathering spot. We are told that visitors could lean against the metal balustrade that surrounded the pool and admire the churning waters. They could stroll inside an enclosure with a tiled roof, Corinthian columns, and a wall decorated with a cheerful fresco of fish and dolphins. Virtually nothing remains of the *castellum* today, but it must have been a popular attraction in its heyday. The lure of flowing water persisted. Many centuries later, between 1768 and 1772, architect Jean Antoine Giral built an aqueduct that brought water to the city of Montpellier, in southern France. He modeled it on Nîmes's Pont

Figure 10.6 *Jean Antoine Giral's water tower at Montpellier, France, built in the classical style. (Imetlin/iStockphoto)*

du Gard, thirty-two miles (fifty-two kilometers) away. The aqueduct flowed into a water tower at the end of the terraced Promenade du Peyrou, which overlooks the city. Giral built Montpellier's tower in the fashionable classical style, complete with imposing columns. Just as in Roman cities, the water tower and its stored water were a matter of civic pride.

Many *castella* were entirely utilitarian. A small *castellum* at Tebourba (Roman Thuburbo Minus), in Tunisia, comprises an oblong chamber, with the aqueduct entering at one end. Three outflow pipes lead from the remaining three sides, one for the baths, another apparently for the theater, a third for the cisterns. Tebourba suffered from water shortages, so the city relied on cisterns rather than continually flowing water. Instead of being a storage tank, this *castellum* was a sort of junction box, where the operators diverted water by using control sluices set into slots on either side of each outlet. One of the sluices controlled two apertures at once and could be set in four different positions, opening both, either one, or neither of them.

There were also secondary *subcastella*, often water towers, where pipes branched off to serve individual households. There were at least 247 *castella* of this type in Rome, but they almost certainly assumed other shapes elsewhere and were sometimes linked in series. According to Frontinus, Rome's authorities often switched the water supply from one aqueduct to another. We know little about the complex interconnections and networks that distributed water in Roman cities, but they must have been of central importance in day-to-day operations. Most public fountains had two water jets drawing on different aqueducts, in case the original supply failed.

Every city water system suffered from cheaters, just like modern-day cable TV providers. Frontinus was very exercised about illegal water usage. He remarked indignantly, "A large number of landed proprietors, past whose fields the aqueducts run, tap the conduits; whence it comes that the public water courses are actually brought to a standstill by private citizens, just to water their gardens."[11] Inside the city, running water was so desirable that many people tapped into the official pipelines, since few households received official authorization for their own supplies. In 11 B.C.E., the Senate passed a law forbidding private users from taking water directly from a conduit. They had to take it from a *castellum*. Until then, there seems to have been little official control over water distribution, to the point that Frontinus complained that the distribution pipes were being "ripped apart" and the flow reduced. A group of wealthy households could collaborate in building and paying for a private line that led from the *castellum* to their own *subcastellum*.

At every *subcastellum*, whether public or private, a standard-sized bronze *calix* (a nozzle) regulated the amount of water that passed into each outflow pipe. This was the only way in which waterworks officials could regulate water supplies to individual households and other users. *Calices* came in various standard sizes, which formed the basis for aquatic bookkeeping, and were fitted to a lead pipe, which had to be the same dimension for a length of 50 feet (15.2 meters), to ensure the correct flow. The opportunities for cheating were open-ended—bribing workers to install a larger size, changing the position of the *calix* in the tank wall, even setting it at an angle. *Calices* not only controlled the distribution and consumption of water but also measured the amount of water handled by each part of the system. Theoretically, the smallest gravity-fed *calix*, 0.9 inches (2.3 centimeters) in diameter, if placed at the same

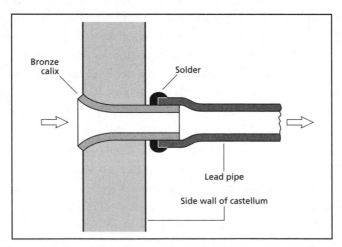

Figure 10.7 *Cross section of a calix. (After A. Trevor Hodge,* Roman Aqueducts and Water Supply. *Copyright © 2002 by Duckworth & Co, London)*

height as the flow full, would have discharged about 10,570 gallons (40 cubic meters) per day. All these figures and our information on *calices* come entirely from Rome, so we have no idea whether similar systems operated in other cities, as seems possible.

The *subcastella* stood at strategic points in the city where they could most easily provide water to surrounding residents. Almost all serviced public fountains either at their foot or some distance away at the end of pipelines. Pompeii offers an excellent example of how a public fountain system worked. Forty water towers, all of them brick structures with lead tanks atop them, are known; at least fifty serviced the city of some 8,000 people, making for about 160 people per fountain. The authorities spaced fountains through the city at intervals of about 109 yards (100 meters). Few people in Pompeii who relied on public supplies had to carry water more than about 55 yards (50 meters). If you allow about 132 gallons (500 liters) per person per day for cooking, washing, and drinking, an arbitrary estimate, the city as a whole needed about 1,056,688 gallons (4,000 cubic meters) a day. Since the city's aqueduct supplied about 1,712,000 gallons (6,480 cubic meters) daily, there was plenty left-over for industrial and public use. Since the bare-minimum allowance for good health is an estimated 2.6 gallons (10 liters) a day, it's clear that Pompeii's public water system was more than adequate for the needs of the populace, disregarding the numerous households who relied on cisterns, wells, and private pipelines.[12]

Rome itself was a city of crowded tenements and high-rise apartment buildings, almost none of which had private water supplies. Their residents, certainly those on the upper floors, had no access to running water in their crowded and unsanitary rooms. They relied on public fountains for domestic water, and on public latrines or chamber pots for sanitation. The fountains themselves comprised a metal spout set at a convenient height, which ran into a stone trough with a drain hole in the base. The trough would normally have been full, so that several people could dip their amphorae into it at the same time. The flow never dried up, so the overflow cascaded over a lip into the street drains, where it helped flush away sewage.

THERE MIGHT HAVE been plenty of public water, but the real priority for aqueducts was the baths. Romans had a passion for bathing. Bathing is part of our daily routine, but this has not always been the case. Standards of personal hygiene were much lower in ancien régime France and in medieval times, when people rarely bathed. This was despite the Christian belief that immersion in water attained peace for one's soul, the ritual of baptism and other rites of passage varying from one sect to another. Bathing was important to the ancient Egyptians, as well as to the Japanese, for ritual purposes. And many Greek shrines centered on sacred pools filled from springs, often with healing properties attributed to a local deity. But the Romans made bathing a near-obsession.

They placed great emphasis on bathing, not only for cleansing themselves but also as a means of entertainment and socializing. Public baths provided water of different temperatures; some, like Aquae Sulis, in southwestern England's modern-day Bath, were prized for their healing qualities. Major baths were palaces of entertainment for all tastes, serving both rich and poor. The custom of regular bathing took hold in Rome during the second century B.C.E. and gradually became a common practice. Small bathhouses, *balneae*, opened throughout the empire, accessible for a small fee. The elaborate *thermae* of later times were the property of the state, but were funded for the most part by the powerful and wealthy, including several emperors. Men and women usually bathed separately or at different times, the women having the worst hours in the morning. They also paid double the charge that men did. Aristocratic women valued their privacy and sometimes owned private steam baths. *Palaestrae*, large gymnasiums for running, weight lifting, and other exercise were part of larger bath complexes. Many *thermae* boasted shops and eating places, as well as libraries and museums, exposing the poor to cultural experiences they would otherwise never have had. The baths were a cacophony. Prominent citizens discussed the issues of the day. Politicians on the make haunted the public spaces. Other clients came to listen to expounding philosophers. The Roman dramatist Seneca wrote of the grunts of weight lifters, the screams from patrons having hair plucked from their bodies. "Then there are the various cries of the pastry cooks, the sausage-sellers, and all the hawkers from the cook-shops, who advertise their wares with a sing-song all of their own."[13]

"Taking the waters," a popular Victorian pastime, was equally popular in Roman times, with notable bath complexes rising at places where natural hot springs came to the surface, as was the case at Aquae Sextiae (Aix-en-Provence) in Gaul. The best-preserved spa is at Bath, where a hot spring bubbled forth at 118 degrees Fahrenheit (48 degrees Celsius).[14] Lead-lined ducting led off the water to fill the Great Bath, about 79 by 39 feet (24 by 12 meters) in area and 5.4 feet (1.7 meters) deep. Heavy lead sheeting also lined the bath. Originally open to the elements or roofed in wood, the pool was later covered by a magnificent concrete barrel vault that was open at both ends. Two smaller, subsidiary pools of different temperatures also received water from the spring. An outflow controlled by a bronze sluice gate carried overflow to the city drains and thence into the nearby Avon River.

Bath was a spa of considerable grandeur, designed to cosset and impress, quite different from more conventional Roman baths, such as those built by the emperor Trajan and others in Rome, and those at such places as Leptis Magna, in North Africa.[15] The latter provided washing facilities and, in colder climates, a place where one could get warm. Above all, they were gathering spots, like the Starbucks of today. People went to the baths to gossip and to socialize, to conduct business and to plot. Baths added significantly to the quality of life in Roman cities, and even military camps, apparently regardless of cost. Major bath complexes centered on a series of large halls—heated by under-floor heating (hypocaust) systems—where people walked and talked. The fountains, plunge baths, and swimming pools were of secondary importance. But whatever the size of the baths, they consumed enormous quantities of water, for the water flowed through them continuously.

Smaller bath complexes such as those at Pompeii and Herculaneum, or at Cyrene in Libya and Abu Mena in Egypt, used simple forms of water-lifting apparatuses operated by slaves, that filled a rooftop tank. Gravity then supplied water to the complex. For this reason, Pompeii had difficulties with bathwater until an aqueduct was built in the first century C.E.—the water table lay eighty-two feet (twenty-five meters) below the surface. The cost of operating one- or two-man lifting devices was high and impracticable for large bath complexes, where an aqueduct was essential. All Rome's great public baths received their water

Figure 10.8 *Interior of the* thermae *of the forum at Pompeii, Italy. (Mimmo Jodice/Corbis)*

from aqueducts, the provision of water being the emperor's responsibility. Sometimes the wealthy built their own aqueducts, and were able to use other sources or existing channels. In 19 B.C.E., the prominent statesman Marcus Vispanius Agrippa built the Aqua Virgo specifically for his private baths. Some baths, like those of the emperor Diocletian, had their own reservoirs, which were larger than any *castellum*. Once the water was inside the bath complex, it flowed through fountains and pools, as well as showers, before draining into the city system. Elaborate lead-pipe networks carried the water into the boiler room, usually housed in a basement corridor under the bath area, where the hypocaust furnace was also located. Trevor Hodge aptly describes such chambers as a "smoky inferno." They cannot have been pleasant workplaces for the slaves who labored there day and night stoking the fires. Once heated, the water entered the bath at about 104 degrees Fahrenheit (40 degrees

Figure 10.9 *Roman public toilet at Ostia Antica, Italy. (Cassianus12/ iStockphoto)*

Celsius), somewhat cooler than at Bath. The hot water and the heat from the hypocausts combined to raise humidity levels high enough to induce profuse sweating. Every bather started in the warm room at about 77 degrees Fahrenheit (25 degrees Celsius), then proceeded to the cold plunge. Once the water passed through the baths, it was put to other uses, often for public toilets for those who did not have private facilities at home.

Public toilets were also the equivalent of a modern-day coffee shop. As many as twenty to forty people would gather happily while conducting their business, sitting on wood or marble seats, often with armrests and curved backs. A constant flow of water from the baths or other sources ran under the latrines and carried off the sewage in short order. A small gutter ran along the floor in front of the seats, so that the occupants could wash their hands. Both patrons of the baths and passersby used the toilets, which usually lay just outside one of the doors adjacent to the street. By no means all public toilets stood alongside bathhouses, but this was certainly a convenient way of using wastewater, which was about the only method of sewage treatment in Roman communities.

ROMAN CITIES, WHATEVER their size, were crowded, making sewage disposal and drainage important concerns, for the comfort level of the inhabitants, if nothing else.[16] But drains for carrying away surplus rainfall, floodwaters, and fountain overflow were considered more important than sewers. The Romans were experts at drain building on large and small scales. They constructed huge canals to drain lakes and marshes, some of them involving the longest tunnels in the ancient world, but their urban drains were often impressive. One of Rome's earliest was the Cloaca Maxima, originally an open channel that carried water from swampland around the Forum into the Tiber. This included rainfall and sewage. It was so large that Agrippa passed through it in a boat. With no U-shaped traps to prevent odors and gases rising from drains, the only solution for Roman engineers was water flow—plenty of it, which carried sewage away as rapidly as possible. This was a major advantage of a water system that depended on perpetual flow rather than water on demand.

Most Roman cities did not have extensive drains or sewers, so drainage flowed at ground level. At Herculaneum and Pompeii, the authorities raised the sidewalks so that pedestrians would walk above the water flowing or trickling down open channels carrying public-fountain overflow. As far as domestic sewage was concerned, many households made use of cesspits, emptied regularly by manure merchants, who collected it and resold the night soil as fertilizer. Many in-house toilets and their cesspits were adjacent to the kitchen; sometimes the pit was in the garden, filled by a pipe from the latrine that was flushed regularly with water. Since the gradient was usually minimal, the odors would have been powerful. High-rise building residents could sometimes rely on large terra-cotta pipes set into the walls, which led to a cesspit, a sewer, or the street. Most of the people crowded into tenements spent the majority of their lives in the open air, made use of public facilities, or simply cast the contents of full chamber pots into the streets, as was also common practice in medieval times.

Even when Rome relied on eleven aqueducts and their copious water, most sewers were open, flowing down the middle of the streets. Emperor Nero is said to have pushed people into them as a joke while walking the streets incognito. Many settlements had nothing more than open sewers, such as, for example, Silchester in Britain, where open ditches nearly two feet (sixty centimeters) wide and two feet (sixty centimeters) deep ran along the streets, carrying away both rainwater and sewage. Silchester prided itself on its state-of-the-art bathing facilities, but the benefits must have been offset by the city's drainage system. Building subsurface drainage systems, especially in long-established cities, was prohibitively expensive and only occasionally undertaken, as it was with Rome's Cloaca Maxima and in Köln (Roman Ara Ubiorum), where over 328 feet (100 meters) of an 8-foot- (2.5-meter-) high vaulted masonry drain with inspection shafts still survives 33 feet (10 meters) under the modern street level.

As for the sewage and wastewater, it usually ended up in a nearby river like the Tiber, or flowed through openings in city walls into smaller channels, as was the case in Athens, from whence the wastewater and sewage fertilized nearby fields and plantations, bringing the hydrological cycle full circle.

AQUEDUCTS WERE EXPENSIVE to maintain, leaked repeatedly, and required constant upkeep. They were profligate ways of supplying water, and could never have been built without copious numbers of prisoners of war and slaves to labor on them. Were they a wasteful luxury? For those who commissioned them, they were not, for these leaders achieved immeasurable prestige from their construction in ways that defined political relationships between the rulers and the ruled. Aqueducts and their imposing arcades were far more than artificial rivers; they were important sources of civic pride, even of Roman identity. However, their existence depended on strong government and the labor of thousands of coerced, anonymous hands. So when the Roman Empire fell apart, the aqueducts that were its pride and joy leaked ever more promiscuously, became clogged with sediment, and eventually ceased to flow. What remained were the traditional water sources: cisterns, wells, springs, rivers, and *qanats*, ways of acquiring water that did not require captives or the enslaved. And these were the ancient water technologies that survived into, and thrived during, the medieval centuries.

PART III

Cisterns and Monsoons

Floods, monsoons, and droughts: In which we look at ancient water management over a broad swath of Asia, where hydrology depends on capricious great rivers and unpredictable monsoon rains. Here, water purifies and is a gift of the gods, the subject of philosophical musings about its uncontrollable qualities.

CHAPTER 11

Waters That Purify

Flashing and whitely gleaming in her mightiness she
Moves along her ample volumes through the realms.
Most active of the active Sindhu unrestrained,
Like a dappled mare, beautiful, fair to see."[1]

THUS THE INDUS RIVER in the Rig Veda—from the Sanskrit *rgveda*,
"praise, verse" and "knowledge"—one of the four primary Hindu sacred
texts, known as the Vedas. The ancient collection of texts in the Rig
Veda were composed in northwestern India between 1700 and 1100
B.C.E., but clearly have roots in earlier times. The ten books, or *manda-
las*, comprise eulogies or prayers used for different sacrificial rituals.
Water features prominently in the Rig Veda, which speaks of a time
when the earth was all water without light:

> *Those waters from the heavens,*
> *Or those waters that flow when dug,*
> *Or even those waters that are self-born,*
> *Flowing towards the ocean, purifying,*
> *May those waters, Oh Goddess,*
> *Protect us here.*[2]

In the Rig Veda, Indra or (Indira) is the supreme god, ruler of thunder and rain and a great warrior. It was he who battled the devil Vrtrá
after he stole all the water in the world. Indra drank copious amounts

of soma, sacred water, passed through the devil's forty-nine fortresses, slew him, and carried water back to earth. It was "he who gave being to the Sun and Morning, who leads the waters. He, O men, is Indira."[3]

The Indus is one of the great rivers of Asia, originating in the Himalayas. Powerful, unpredictable, and sometimes violent, its waters flow through arid landscapes for thousands of miles before emptying into the Arabian Sea. During winter, ice dams and landslides block the river deep in the Himalayas. When the snow melts, the impounded water rushes downstream, sometimes with catastrophic force. The Indus itself and five other rivers converge near the modern city of Multan, forming the lower river. For thousands of years, these rivers' floodplains sustained farming populations, long before the great cities of Harappa and Mohenjodaro came into being.

A second river, anciently known as the Sarasvati, now the Ghaggar-Hakra, once flowed through the Great Indian Desert. This one is also praised in the Rig Veda, as one of the seven river sisters of the region: "Foremost of mothers, foremost of rivers of goddesses, Sarasvati . . . In thee, Sarasvati, divine, all generations have their stars."[4] Sarasvati apparently never reached the Indian Ocean during Harappan times, for it terminated in an inland delta naturally irrigated by annual floods. Eventually, the river dried up. The Indus, the Sarasvati, and their tributaries nurtured early Indian civilization in an arid environment of extreme heat and insufficient rainfall, where famine was always hovering in the wings. Each summer, the floods brought both water and silt that fertilized floodplain soils for hundreds of villages, each with their own local shrines and close ritual ties to the life-giving rivers.

A TRADITION OF self-sufficiency, and clever water management, went back to the beginnings of Indian civilization in the northwest, in what is now Pakistan, and even earlier, to the ancient farming villages that first cultivated the soils of the Indus River. Between 2500 and 1900 B.C.E., these traditions of water harvesting culminated in the Harappan civilization, named after the city of Harappa, on the Ravi River. Harappan civilization was a network of cities, towns, and villages covering an area

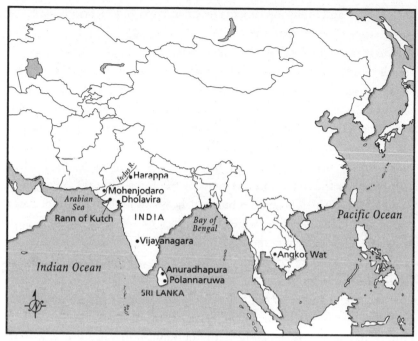

Figure 11.1 *Map showing locations in chapter 11.*

of three hundred thousand square miles (seven hundred thousand square kilometers). As was the case in so many other places, agriculture was a local matter, administered at the village level.[5] There were no bombastic rulers boasting of their achievements on palace walls. Harappan leaders were unostentatious people, with none of the ardent militarism of the Assyrian kings or the self-glorification of the Egyptian pharaohs. We know little about the Harappans, except that they were efficient urban administrators, who developed sophisticated methods of water management and sewage disposal for their cities that were without peer at the time.

Harappan civilization had much earlier roots, but reached maturity with dramatic rapidity, with the founding of the city of Mohenjodaro on the Indus River in about 2450 B.C.E. A period of explosive growth transformed life along the Indus and its tributaries, for reasons that are little understood but were probably connected with long-term changes in river flow and monsoon rains. The great river and its floods were the cornerstone of Harappan life.

The Harappans lived in a world of extremes: hot, dry summers, cold and dry winters. As in other such environments, the amount of precipitation varied, often dramatically, from year to year, as it does today. Climatic conditions have not changed since then. Rainfall came from two directions. In winter, storms from the Mediterranean area terminated in northwestern India, usually to the north of the Indus region. More important, the summer months brought monsoon rains, pushed northward by the seasonal movement of the Intertropical Convergence Zone (ITCZ), which brought a burst of heavy rainfall in its train. River floods and monsoon rains were the sources of Harappan water, both markedly seasonal phenomena. For this reason, careful harvesting of groundwater and the flood and, above all, meticulous storage of water for the long dry months were of paramount importance in Harappan life. Irregular, capricious rainfall made water storage in tanks a central part of South Asian life from the earliest times.[6] Wells, too, played an important role in a society where an abundance of water was essential, not only for daily use but for ritual purposes as well.

Why, then, did the Harappan civilization suddenly breed cities? The answer may lie in the movements of the Intertropical Convergence Zone. The position of the ITCZ has changed through time. For much of the past ten thousand years, it has moved north rapidly in late spring and early summer. On at least four occasions, global bursts of volcanic activity have suppressed the northward movement, bringing less rainfall to the Indus region. By modeling the position of the ITCZ, a team of researchers has calculated the discharge of the Beas River, near Harappa.[7] Between 3600 and 2100 B.C.E., the river's discharge rate increased sharply, by about 27 percent over almost 1,500 years. In the lower Indus, the increase may have been as much as 300 percent. This period of increased river flow coincides with the mature Harappan civilization and the rapid growth of cities like Harappa and Mohenjodaro. After 2100 B.C.E., the movement of the ITCZ slowed, rivers received less water, and droughts were more commonplace. Two hundred years later, the Indus cities were largely deserted.

OF THE FIVE major Harappan cities, we know most about Mohenjo-daro, on the lower Indus, with an estimated thirty-five thousand to forty thousand inhabitants and six times the area of Harappa, on the Ravi.[8] Mohenjodaro was a city of at least partially planned streets, with a conspicuous citadel at the western end protected by fortifications and flood works. There were no spectacular shrines atop the citadel, just public structures, which included a pillared hall and a large bath that most likely had ritual associations, though of course, we cannot be sure. The priests and rulers who dwelled on the summit looked down on a crowded city of narrow streets and alleyways, houses laid out around courtyards, bazaars, and artisans' quarters. The urban precincts lay on artificial platforms above the high flood level of the nearby river. The inundations were a major concern, for rising village populations on the floodplain had stripped the forest and grass cover that had prevented soil erosion and provided natural controls for raging floodwaters. In wet years, such as were common after 3000 B.C.E., the Indus floods would sweep everything before them. Confronted with what may have seemed like the wrath of the gods, the people had only one defense: elaborate flood works. Irrigation agriculture, with its canals, drainage channels, and storage tanks, not only fed more mouths but also provided at least a degree of security from the moods of the major rivers.

The crops grown by the Harappans included wheat; barley, grown in the winter; and millet, a drought-resistant summer crop. With these staples and a wide diversity of fruit and minor crops, the farmers practiced multicropping (growing two or more crops in one field), using humped cattle to pull their plows, for herding was an important part of the subsistence economy. The irrigation methods used on the Indus were, of course, gravity based, but are little known, all traces of the ancient systems having long since vanished. We can assume, however, that they were broadly similar to those used elsewhere in the ancient world at the time. Where the Harappans stood out was in their lavish use of water in their cities, which came not only from tanks but also from intensive harvesting of groundwater. Generations of excavations at both Harappa and Mohenjodaro have revealed water-management techniques that were unrivaled anywhere until Greek and Roman times.

Before constructing any infrastructure, streets, or dwellings, the

Mohenjodaro authorities invested enormous efforts in building up
fired-brick platforms to raise the city above the Indus floodplain. One
estimate calculates that twenty-five hundred laborers would have taken
about four years and four months just to construct the foundations of
the city. Even with this expenditure, the area of the community would
have been relatively small, which suggests that Mohenjodaro's leaders
had a long-term plan for it. (There may have been outlying communities,
but they were vulnerable to floods.) Confined as it was to the original
brick platforms, central Mohenjodaro rose vertically over the generations, with continual rebuilding and platform construction. This meant
that water management presented unique challenges, especially as the
water table was ever deeper below the city foundations, inaccessible except with wells . . .

The circular hole lies in the middle of a brick platform spattered
with clay and mud. A strong leather bucket holds a clutch of wedge-
shaped bricks at the edge of the well. Two men lift the heavy container
into the well as two others bear the strain on the fiber rope. They call
into the dark pit; a voice echoes an affirmative from far below. The
bricks vanish from sight and descend smoothly at the end of the rope.
Abruptly, the cord slackens; the men relax, then haul up a load of clay.
Far below them, a man toils alone in the dark, hacking away at the

Figure 11.2 *A public well at Mohenjodaro, Pakistan. (Mark Kenoyer/*
Images of Asia)

circular walls, ankle deep in mud and sand. As each bucket is filled, he carefully lays another course of bricks to line the wall . . .

At least seven hundred wells brought water into the heart of the city, all of them carefully lined with wedge-shaped fired bricks. Brick-lined wells were a Harappan invention, such vertical water-supply systems being virtually unknown in contemporary Egyptian and Mesopotamian cities. The bricks supported the earth walls and prevented erosion at the surface. They also filtered the water and kept out silt and other contaminants. Mohenjodaro's wells were of all sizes, some as small as 2 feet (around 0.6 meter) across. The average was about 3 feet (1 meter), the largest being 6.8 feet (2.1 meters) in diameter. Surprisingly, most of the wells were originally dug when the city was first founded, then additional brick courses were added as needed. Some were abandoned, but no new ones were commissioned. Today, the exposed specimens rise like brick chimneys from archaeological trenches. Plotting the wells on a city map, one finds that they were relatively evenly spaced at about 118-foot (36-meter) intervals, each serving an average area of 17,270 square feet (1,326 square meters), with an average of one to every third house. Some wells were private, located in houses and inaccessible from the street. But most of them were apparently for public use, sited in paved rooms whose floors were deeply worn by generations of heavy water jars. In contrast, in ancient Egyptian and Mesopotamian cities, most families collected water from the river and stored it in cisterns at home or used it immediately.

Archaeologist Michael Jansen believes that the Harappans may have developed wells as a result of their long experience with extracting potting clay from deep pits.[9] They obviously knew that a circular well was the structure most resistant to soil pressure at depth. Wells gave the city's inhabitants self-sufficiency in an arid climate, for the Indus flowed unpredictably past Mohenjodaro. As Jansen points out, seven hundred wells means that the inhabitants of this and other Harappan cities consumed vast quantities of water, especially for bathing, far more than would have been required merely for personal hygiene. Thus, the liberal use of water by a people whose farming depended on seasonal inundations can only have had a deeper spiritual meaning, reflected in the great bath at the high point of the city, the citadel.

The Great Bath of Mohenjodaro, a pool 39 by 22.9 feet (12 by 7 meters) in area and 7.8 feet (2.4 meters) deep, received its water from a double-walled, cylindrical well in one of the rooms by the pillared gallery associated with the bath. Specially manufactured bricks formed the inner shell of the well, laid with gypsum mortar and sealed with a bitumen layer between this and the outer brick layer. As far as one can tell, the bath was filled by hand, a laborious task to say the least. The used water flowed out through an outlet at the southwest corner, into a large corbel-roofed drain. Just what the Great Bath was used for is unknown, but almost everyone agrees that it was a ritual pool, presumably used for rites of purification.

Bathing is an integral part of Indian life and ritual practice to this day, both in private acts of worship and in great public rituals. Cleansing with water is a sacred obligation. Millions of Hindu pilgrims immerse themselves at dawn in the waters of the Ganges River at places like the temple town of Haridwar as part of the Maha Kumbh Mela ritual, which takes place every thirteen years. They flock to bathing places at the foot of a shrine of the high god, Lord Shiva, to pray to the sun as it rises over the nearby hills. Vedic legends state that Haridwar was one of the four places on earth where a drop of the divine nectar of immortality fell from the pitcher that Garuda, the divine bird of the

Figure 11.3 *The bath on the citadel at Mohenjodaro, Pakistan. (Mark Kenoyer/Images of Asia)*

supreme deity, Lord Vishnu, spirited away from the demons during the war between them and the gods. Some archaeologists have claimed that Harappan seals bear images of an early form of Shiva, but this notion is much disputed. There is no doubt, however, that some of today's Hindu beliefs and rituals have ties to much earlier beliefs, perhaps some of those connected to Mohenjodaro's imposing bath complex.

Whatever its use, the Great Bath depended on a vertical water system, not aqueducts like those constructed by the Greeks and Romans. Maintaining the Mohenjodaro bath required a great deal of manual labor, far more than would have been required for domestic water supplies, suggesting that this magnificent structure had unusual significance in Harappan life.

Mohenjodaro and other Harappan settlements boasted remarkably efficient and well-designed drainage systems. Every street and lane in Mohenjodaro possessed U-shaped, brick-lined drains, which ran along one side of the unpaved defile. They flowed at a gradient of about 0.8 inches (2 centimeters) per 0.6 feet (1 meter), with closed cesspits to catch solids located wherever drains met. The pits were vertical shafts, accessible for cleaning as the sediment in them accumulated; open soak pits were also common, especially at intersections between smaller and larger streets.

The sedulous attention paid to water supplies and drainage in an environment where less than 5 inches (127 millimeters) of rain fell annually is quite remarkable. It may be connected with the apparent preoccupation with bathing and, presumably, ritual cleanliness. Most of the water that flowed through the drainage system came from bathing platforms, built into the street side of many houses. Each was about 4.9 feet (1.5 meters) square, with a carefully prepared brick floor surrounded by a low rim to form a shallow basin, slightly tilted so that the dirty water flowed to a corner, then out of the building through a drain into either a soak pit or the street drain. Often, a short flight of stairs led to a small platform, where a family member or a servant could pour water on a bather in an early form of the shower.

Latrines, often located on the outer wall of bathing platforms, came in several forms, among them the simple hole in the ground located above a cesspit or a vertical chute in the wall of a house, which emptied into a cesspit or a street drain. Some latrines allowed users to sit down on a wooden plank or a brick platform. The effluent and bathing water sometimes passed into closed sewage-catchment vessels located under the toilet chute and tipped out regularly. Some houses had catchments set into their walls, which had to be scooped out frequently. In some cases, there was a gap between the outlet and the drain. The sewage smell throughout the city would have been overpowering unless a considerable water flow and constant maintenance cleaned out the drains several times daily. Such cleansing would have required large numbers of sanitation workers. Whether these worked for individual families, groups of households, or the city as a whole is unknown.

The urban infrastructure at Mohenjodaro remained unsurpassed until classical times, especially in its even distribution of water and drainage facilities across the city, apparently with more-or-less equal access for all. Why this was done and what kind of society lay behind these innovations remains a complete mystery, but it seems likely that powerful spiritual obligations helped shape society and its water distribution.

WE KNOW VIRTUALLY nothing about the tanks that must have surrounded Harappa and Mohenjodaro, but another city, Dholavira, located in a much drier environment, depended entirely on reservoirs and seasonal rainfall.[10] Dholavira lies on Khadir Island, in Gujarat, India, surrounded by the arid wastes of the Great Rann of Kachchh (Kutch). Fourteen seasons of excavations by the Archaeological Survey of India have uncovered a city occupied for over fifteen hundred years, through the third millennium B.C.E. and into the second. Dholavira is shaped approximately like a parallelogram, with fortifications that boasted seventeen gates, a citadel, and an elaborate system for harvesting water in a harsh environment on the edge of the monsoon belt. There are no perennial lakes, rivers, or springs here; monsoon failures several years running are not uncommon. The groundwater is mostly brackish and unfit for animal and human consumption, even for agriculture. Dholavira's inhabitants depended almost entirely on irregular monsoon runoff captured in storage reservoirs.

The founders of the city chose the site with care, laying it out on sloping terrain that dropped about forty-three feet (thirteen meters) from east to west within the walls. They built the natural gradient into their layout. Two monsoon storm channels embrace Dholavira. The Manhar channel, flowing in from the east and running down the south side of the settlement, was the first channel to be dammed and deepened, forming a large reservoir east and south of what began as a fortress and became a city. The other monsoon channel, Mansar, lies on the northern side of Dholavira. From the beginning, reservoirs formed an integral part of the city, sixteen of them, of different sizes, arranged in series on all four sides. About forty-seven acres (ten hectares), some 10 percent of the total area inside the outer fortifications, was devoted to reservoirs and other water structures. The natural gradient allowed for cascading series of reservoirs, separated from one another by massive, broad embankments but linked with feeding drains.

The Manhar fed what is the largest of the reservoirs excavated so far, which measures 241 by 96 feet (73.4 by 29.3 meters) and is surrounded by an embankment at least 4 feet (1.2 meters) high. The tank has three levels dug into the alluvium and bedrock, the deepest 34.7 feet (10.4 meters) and accessed, apparently by the inhabitants at large, by stairways at three of the four corners, the fourth being devoted to a waste weir. Five more reservoirs lie south of the citadel, two of large size forming one unit, cut into the rock with basins of varying size and access stairways. Drains connect each tank with the others, thereby forming a series. The embankments that separate them served not only to reduce pressure on the reservoir walls but also as walkways. These bunds also acted as repositories for the silt removed from the reservoirs during periodic cleaning. A conservative estimate calculates that Dholavira's tanks would hold at least 79,250,000 gallons (300,000 cubic meters) of water, this without adding other water supplies in reservoirs outside the walls. Even with this capacity, there was sometimes cause for concern, so additional precautions came into play. The floors of many reservoirs featured deep depressions or troughs where water would remain in exceptionally dry years.

There was nothing especially complex about Dholavira's water management, which relied on a technology of dam building, albeit much refined, that had already been in use in mountainous Baluchistan to the northwest centuries earlier, to block gaps in hills and trap storm water from higher ground. Dholavira even collected rainwater that fell on the citadel with a complex drain network. Some of the larger drains were high and broad enough to allow one to walk through them. None of these drains carried sewage, just monsoon runoff, which passed through cut-stone cascades into a tank carved out in the western part of the citadel. A large, very deep well, which would have had to reach a depth of seventy-nine feet (twenty-three meters) to hit groundwater, lay in the southwestern part of the citadel. The water from the well, perhaps lifted in a large leather bag, fed two tanks, whose significance is unknown. There are few wells in the city, which is hardly surprising in view of the unpalatable groundwater. However, Dholavira maintained a well-organized sewage system that recalls those at other Harappan cities.

AT DHOLAVIRA, WATER management depended on interconnected series of reservoirs receiving water from up-gradient and monsoon runoff. These were quite different from reservoirs fed by perennial rivers or canals, and from isolated tanks that depended on a single catchment, which watered gravity-driven irrigation systems over much of southern India for many centuries. Such water management worked well, even with the unpredictable and often localized monsoon rainfall of South India. Vijayanagar, one of the largest cities of ancient South Asia, flourished on such water systems for nearly six hundred years, and came into being in the early 1300s C.E., as a shrine on the south side of the Tungabhadra River, a location that gave legitimacy to its ambitious rulers.[11] The walled city depended on an agricultural zone that extended over 135 square miles (350 square kilometers), a landscape of small towns and villages near the river. A series of canals and a large canal-fed reservoir, known as the Kamalapuram *kere*, supported wet rice cultivation. At the same time, other reservoirs on the north side of the Tungabhadra stored monsoon runoff for dry agriculture.

A second phase of expansion and rural and urban population growth occurred at Vijayanagar during the early sixteenth century. This time, the growth was spectacular: The landscape filled in with small villages, imposing monumental buildings rose in the city, and settlement spilled outside the city walls. Almost all agricultural production now focused on rice, orchard crops, and vegetables for the city. Villagers grew cotton in dry fields, on a very large scale indeed, for the local textile industry, while large herds provided meat for the city's inhabitants. Now water management changed, with the construction of the large Agedondi canal, fed by the Tungabhadra River, and an aqueduct north of the river, which allowed the expansion of irrigated rice cultivation. The increasing scale of irrigation transformed the landscape. Long, complex chains of reservoirs extended into the Daroji Valley, south of the Tungabhadra. Terraces and check dams helped reshape the terrain. Modifying slopes altered drainage patterns and water movement. Many of the irrigation canals remained in use until modern times, long after the defeat of Vijayanagar's army and the burning of the city in 1565 C.E. The greatest Daroji reservoir is still in use, largely because strenuous efforts have kept it well maintained. However, most of Vijayanagar's reservoirs, especially those constructed during the great agrarian expansion, are now buried deep under silt or have had their embankments breached.

VIJAYANAGAR'S CANALS AND tanks pale into insignificance alongside those developed by Sinhalese rulers in Sri Lanka at Anuradhapura and Polonnaruwa. Anuradhapura, in the northern part of the island, began life as a small agricultural village with circular timber structures before the fourth century B.C.E. and grew to become a great medieval city with trading contacts over an enormous area of the Indian Ocean world.[12] The city was abandoned in 1017 C.E. Nine and a half square miles (twenty-six square kilometers) of Buddhist monastic establishments surround what was the secular core of the city, centered on four large brick-built stupas, hemispherical mounds that served as places of veneration. They formed a sacred complex, which came into being during the third century B.C.E. with the establishment of the monastery by King Devanampiya Tissa. Pilgrims from every part of the Buddhist world visited this wealthy and prestigious religious center, with its five thousand to eight thousand monks. The communal efforts of thousands of devotees built

Anuradhapura's monuments, supported by enormous food surpluses and labor available to its rulers, much of it derived from irrigation agriculture.

Three artificial lakes, built between the fourth century B.C.E. and the first century C.E., define the outer limits of the city. They cover between 225 and 3,183 acres (91 and 1,288 hectares) and were augmented with feeder channels and canals in the fifth century C.E. The rulers stored excess water from the wet season for drinking and agriculture. At the same time, they diverted water from other river catchments to large storage tanks, which could be released into Anuradhapura's system.

A five-year research project involving Sri Lankan archaeologists and teams from British universities is mapping nonurban communities, as well as the soils and sedimentary geology, of Anuradhapura's hinterland.[13] The survey has yielded over one thousand sites, among them settlements and manufactories, as well as 235 landscape features, including tanks that supported both nearby villages and cultivation. These were associated with anicuts (small irrigation dams), sluices, and irrigation channels. Sixty-eight sites were what the researchers call "monastic" locations,

Figure 11.4 *Anuradhapura, Sri Lanka, seen across the Bassawak Kulama Tank. (Michael Freeman/Corbis)*

sacred places associated with Buddhist inscriptions and various monuments. Thirty of these were rock shelters; others, platforms located
within Anuradhapura's monastic area. One royal monastery, Pathigala,
was only accessible by dugout canoe during the wet season. The monasteries in the hinterland were linked to the great monasteries of Anuradhapura itself, perhaps through sight lines that connected their stupas to
those in the distant city. Thus, the main ritual centers formed a single
whole. While excavating Vehragala, an abandoned monastic site on a
conspicuous outcrop, archaeologists were able to observe ancient monastic functions still under way in the area—labor, agricultural surpluses, money, building materials, knowledge, and religious merit were
redistributed by the monks to incumbents, villagers, and visiting pilgrims. The monasteries imposed, and still impose, an ideological control over an irrigated hinterland, creating a theocratic landscape, where
monks served as both religious and secular administrators.

Polonnaruwa, in east-central Sri Lanka, is the second-most ancient
of the island's kingdoms, founded by Sinhalese king Vijayabahu I in
1070 C.E.[14] Being further inland than Anuradhapura, the site offered
strategic advantages. Vijayabahu's grandson, Parakramabahu I (1153–86),
brought a prosperous era of trade and agriculture to the city, vowing
that not a drop of water would be wasted. He commissioned canals
and tanks even more extensive than those at Anuradhapura, including
the Parakrama Samudraya, or the "Sea of Parakrama," which encircles
the city and served both as a defense against attackers and as the major
water source. The sea has a surface area of 8.72 square miles (22.6
square kilometers) and a catchment area of 29 square miles (75 square
kilometers), drawing water from small streams and a channel from a
river nearly 5 miles (8 kilometers) to the north. It is, in fact, three tanks
connected by narrow channels at low water, the oldest having been built
in about 386 C.E. Parakramabahu added two more, with the highest
one at the eastern end, enclosing the reservoir with a bund that is forty
feet (twelve meters) high and forms an 8.5-mile (13.7-kilometer) peri-
meter. Thousands of laborers took years to excavate the tanks and
form the reservoir. We can imagine the rows of villagers, clad in loincloths, up to their knees and thighs in mud, laboriously digging out the
heavy soil and hefting it from hand to hand atop the high embankments

that block their view on all sides. High above them, robed priests watch silently. The men work willingly, for their reward is a spiritual one. The enormous labor of digging the sea was no luxury, for the artificial lake nourished an elaborate rice-irrigation system that covered eighteen thousand acres (seventy-three hundred hectares) and supported a dense urban population. Parakrama Samudraya and its canals fell into disuse after the abandonment of the city in the fourteenth century, but the reservoir has now been restored and once again supports rice cultivation.

As was the case in Bali, local temples and monasteries played critical roles in water management, agricultural production, and conflict resolution within their immediate hinterlands. Monasteries were focal points for the great religious festivals of the year, attended by thousands of people, urban and rural dwellers alike, like the Jasmine Flower Festival of today. One is irresistibly reminded of the great Maya ceremonies performed in the shadows of tall pyramids, which served to link the rulers and their complex water rituals with their subjects in a durable social contract. But the waterworks of the Maya and the Sinhalese were minuscule compared with those of the ancient Khmer of Southeast Asia.

ANGKOR WAT, IN Cambodia, is one of the architectural masterpieces of antiquity. The Khmer king Suryavarman II began his extraordinary shrine in 1113 C.E. Angkor Wat reproduces part of the heavenly world in a terrestrial mode in a monument to the Hindu god Vishnu, the preserver of the universe.[15] This is the largest religious building in the world, dwarfing the greatest Sumerian ziggurat and making Mohenjodaro's citadel look like a village shrine. The central block towers more than two hundred feet (sixty meters) above the forest. Superbly executed bas-reliefs show Suryavarman receiving high officials, progressing down a hillside on an elephant. The court rides with him through a forest, with noble ladies in litters. Scenes of battles and celestial maidens appear throughout Angkor Wat. Naked to the waist, slender and sinuous, the girls wear skirts of rich fabric. Here, the king communicated with the gods. When he died, he was buried in the central tower, so that his soul made contact with the royal ancestors and he became as one with Vishnu. As many as 750,000 people lived in the greater Angkor area,

Figure 11.5 *Angkor Wat, Cambodia. (Olga Anourina/iStockphoto)*

which encompassed Angkor Wat and the city of Angkor Thom. Forty thousand of them lived in the 3.5 square miles (9 square kilometers) inside the walls of Angkor Thom. The largest urban complex of the ancient world, Angkor Thom covered an area equivalent to New York City's five boroughs.

How did the Khmer manage their water supplies to grow crops and support a large urban population? For the past ten years or more, Australian, Cambodian, and French scholars have mapped much of the hinterland surrounding the temple, with surprising results.[16] They've used not only aerial photographs but also satellite images, producing a map covering about 1,158 square miles (3,000 square kilometers) around Angkor Wat. We now know that Angkor developed a huge network of channels, embankments, and reservoirs that covered over 386 square miles (1,000 square kilometers). From the air, the artificial landscape of canals and reservoirs unfolds like a huge jigsaw puzzle of tanks and shrines, dispersed hamlets and long-abandoned embankments, a hinterland that is as much water as forest and open country. High above the plains, your eye is drawn irresistibly to the center, to Angkor Wat and its *barays* (reservoirs). This is a world with a center and a periphery. You get a strong impression of water as an overwhelmingly powerful social force at Angkor, more so than almost anywhere else in the ancient world.

Angkor Wat lies in an environment where seasonal monsoon rainfall is insufficient to support large-scale rice production. Monsoons between May and October bring rain gushing down from the Kulen Hills. The water spreads rapidly across the gently sloping plain and eventually into the Tonle Sap, the Mekong River–fed basin of central Cambodia some ten miles (sixteen kilometers) to the south. The Khmer built enormous *barays* to store water for the dry season and drought years. The population surrounding the royal precincts lived in dispersed hamlets along embankments and close to temples and shrines. Everywhere, a linear network of channels and embankments conformed with the random distribution of local shrines, water tanks, and occupation mounds. The system combined channels with clay-and-sand embankments, which slowed and dispersed incoming floodwaters across the landscape during monsoon season. Much of the water ended up in reservoirs, the largest of which, the West Baray, is about 5 miles (8 kilometers) long and 1.2 miles (2 kilometers) wide, fed today by water diverted from the Siem Reap River. Careful estimates suggest that at least two hundred thousand laborers may have dug the *baray* and piled soil onto the surrounding embankments. The sheer scale of the waterworks, built entirely by hand, beggars the imagination. The West Baray alone held about 13,210,000 gallons (50 million cubic meters), while the embankments were about 394 feet (120 meters) wide and 32.8 feet (10 meters) high. One linear embankment followed a shallow channel that begins near the southwest corner of the West Baray and extends at least 25 miles (40 kilometers). Embankments steered the water across the landscape and in the flood season served as barriers. They were also roadways, so the entire network had a flexibility in terms of functions, which varied with flood levels and the time of year.

Looking at Angkor Wat from the air and from space has allowed researchers to examine the complex water system as a whole. Three large *barays*, including the giant West Baray, were at the center. To the north, between the Kulen Hills and the West Baray, was a collector-and-flow management system of channels and embankments that spread water

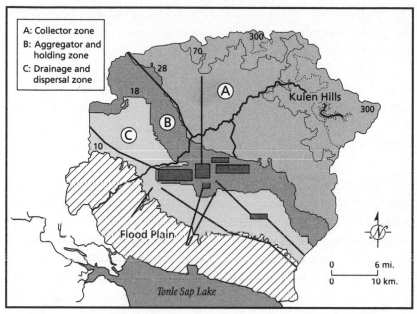

Figure 11.6 *Angkor Wat's hinterland. (After Roland Fletcher et al.,* "The Water Management Network of Angkor, Cambodia," Antiquity *82, 4 [2008]: 658–70, fig. 1. Courtesy: Antiquity Trust)*

and directed it southward. The so-called Great North Channel ran for 15.5 miles (25.5 kilometers), from the northern hills to the north gate of Angkor Wat. The system formed a complex grid of long, linear embankments and channels aligned on the cardinal directions, with numerous right-angle turns and cross-channels that affected water flow and the sedimentation rate, of critical importance in this landscape of low relief and sediment-bearing water. The managers had many options for directing water, including zigzag canals and at least one masonry spillway built of interlocking blocks with a gradient of about 2.5 percent, which is about 164 feet (50 meters) wide and about 295 feet (90 meters) long. This imposing structure, part spillway and part dam, served to divert water when needed into a long canal that altered the course of the Siem Reap River, redirecting it toward the East Baray. Angkor's engineers were fully capable of building permanent water-management structures that enabled them to handle excess water by diverting it during the monsoon. Many details of this complex system still await clarification.

The central zone of the system comprised both the major *barays* and the extensive temple moats, built between the ninth and thirteenth centuries, the latter symbolizing the oceans that surrounded the home of the gods. Enormous quantities of water flowed into the core. The intake channel for the West Baray alone was about eighty-two feet (twenty-five meters) across. Still little-explored smaller channels allowed water to be diverted from the *baray* to the eastern side of the temple and elsewhere. At the same time, water soakage from the *barays* combined with surface evaporation in the surrounding fields would have drawn up groundwater to nourish growing crops. By growing rice in enclosed fields, the Khmer managed not only to grow ample food but also to control flooding, while having enough water to survive in dry years. To the south was the dispersal system, channels that both distributed water and carried it south and east out of the central zone. The East Baray had a massive exit channel that diverted water to the east and southeast. The many channels to the south were straight, clearly intended to move water away as rapidly as possible.

Centuries of these large-scale waterworks completely transformed the Angkor landscape. The entire system covered a terrain with a shallow gradient variation of 0.1 percent, which meant that water distribution and flow required constant adjustment to maintain stable water levels and even flow. Silting was a long-term problem. There is reason to believe that by the twelfth century the water engineers were attempting to combat pervasive silting by digging channels that helped them dispose of excessive water and thereby decrease sedimentation rates.

Angkor Wat was the heart of a sacred landscape, for here, as in Bali, water management and ritual were closely intertwined. When the Khmer built the West Baray, during the eleventh century, they also erected an exquisite water court, the West Mebon, where a 19.6-foot- (6-meter-) long figure of Vishnu reclined. The temple reminds us that the gods watched over the anonymous laborers who toiled over *barays*, channels, and embankments.

In the end, the constantly modified system likely became difficult to control. Routine maintenance alone must have consumed the labors of large numbers of people, especially when major floods overwhelmed dams or embankments. Everything depended on a reliable monsoon. If several monsoons failed—and they provided 90 percent of Angkor's water—then the kingdom was in trouble. Rings from long-lived *po mu* cypress trees growing in Vietnam record major droughts that lasted from 1362 to 1392 and from 1415 to 1440, periods when the monsoons either failed or were anemic at best.[17] These dry spells, caused by persistent El Niños, occurred during the early part of the Little Ice Age, a period of unsettled climate and extreme weather, and a time when, pollens show us, the West Baray supported marsh and terrestrial plants, something quite different from the aquatic plants of earlier centuries. Yet, ironically, while the *barays* may have dried up, the Mekong River, fed by distant Tibetan glaciers, still had plenty of water when the monsoons failed. Khmer engineers were expert water managers, but relying as they did on gravity as their only pump, they could not move the Mekong's waters upstream.

In 1431, Angkor collapsed, perhaps at the hands of the kingdom of Ayutthaya to the west. The conquering ruler installed his son at Angkor, but the center of economic gravity was shifting downstream, away from the Tonle Sap and close to the Mekong River, which provided easier access to the trade routes of the South China Sea. However, the droughts recorded in the *po mu* hint that the dense population of Angkor exceeded the ability of the land, and that of their complex, silt-threatened water system, to support them. It may be no coincidence that Ayutthaya's invasion occurred as the second great drought was ending. A short time after the conquest, much of the six-century-old water system imploded, to be largely forgotten until today. Like the Sassanians of Mesopotamia, who depended slavishly on irrigation and its yields, and the Maya of Central America, described in chapter 13, who ravaged a tropical rainforest environment, Angkor delivers a powerful message about the dangers of overstressing the environment, and about maintaining sustainability, a message we would be well advised to heed.

CHAPTER 12

China's Sorrow

WEI RIVER, NORTHERN CHINA: 2270 B.C.E., late summer. The laborers have traveled many miles upstream to reach their inhospitable defile. Each year, they return to repair gaps in the weir, torn by the Wei's summer flood. Practically naked in the heat, some laboriously split rocks and heft them through gravel and mud to the water's edge. More villagers maneuver the fiber nets full of stone into the fast-moving stream, up to their waists in the current for much of the day. A man slips and vanishes under water. His companions keep struggling with the heavy mesh. They dare not let go for fear of losing their load. Fortunately, his feet land on a gravel bed, and he struggles to his knees, then staggers back upstream. The work just keeps going, an unending routine of hammering and lifting, of battling inexorable water. There's nowhere to rest, little shade, always yet more long tree trunks to be cut to size and hammered in place to stabilize the bags of stone that make up the weir. An imposing, mustachioed figure stands impassively atop a nearby dike accompanied by soldiers armed to the teeth. The workers bend to their labor under his critical eye, for Yu is a feared, fierce taskmaster. Rations are meager, every move watched by supervising officials. Dozens of workers perish each year, for as quickly as the laborers rebuild the dam, equally swiftly the river breaches it once more . . .

Yu the Great, an iconic figure in Chinese history, adorns a jade boulder found near Hotan, in the far west of the imperial domains, in the late 1770s C.E. A hundred horses assisted by one thousand laborers dragged the precious find to Beijing, a task that took three years. Emperor Qianlong himself chose the subject to be carved on the 11,750-pound

(5,330-kilogram) stone. He selected a painting by an anonymous Song Dynasty artist of centuries earlier, which depicted Yu the Great taming floodwaters in the remote past. The carving, which still stands in its original Beijing location, took three years to carve and a fourth to inscribe with a eulogy to Yu and the emperor.

Yu (said to have lived from 2297 to 2198 B.C.E.) was an alleged descendant of the legendary warlord Huang Di (the Yellow Emperor), mythic founder of Chinese civilization as early as 2698 B.C.E.[1] Yu was a skilled warrior and the founder of the Xia Dynasty. But his lasting fame comes from his prowess at controlling river floods not by containing them but by using channels and ditches to steer them away from farmland. He is said to have spent thirteen years away from home trying to control the surging waters with canals rather than dams. After thirty years, and the loss of many lives, he apparently succeeded and took a lead in developing agriculture.

The Xia Dynasty ruled one of the kingdoms that formed the primordial Shang civilization, centered on the Huang He (Yellow River) Valley of northern China. Yu's flood-control efforts must have been transitory, for later Shang rulers repeatedly moved their capital because of the menacing river. A Ming Dynasty commentator remarked many centuries later that "the shifts of the Yellow River seem to be guided by some god, and not to be something in which human efforts can be involved."[2] For four thousand years, the Chinese attempted to master the Huang River by encouraging its waters to flow out to the China Sea.

China has two major river systems, which have shaped much of its history. In the north, the Huang River, often called the mother of Chinese civilization, and perhaps more appropriately, "China's Sorrow," flows through arid terrain and a basin that covers more than 290,500 square miles (742,443 square kilometers). Like the Nile, it floods each spring and carries both irrigation water and a heavy load of brown-yellow silt, whence its common name. The northern lands, with their harsh, often water-deprived environments, were one pole of the ancient Chinese world, the other lying to the south in warmer, better-watered landscapes. The Yangtze, in southern China, is the longest river in Asia and flows through milder environments, where rice cultivation has thrived since at least 4000 B.C.E. With these two river systems as central

Chinese Dynasties

This simple listing, purely for general guidance, masks great historical complexity and overlaps between rulers and dynasties.

Northern Chinese states

? 2200 B.C.E.	Xia
1766–1122	Shang
1122–475	Zhou
475–221 B.C.E.	Warring States

Dynasties

221–207 B.C.E.	Qin (unification of China)
206 B.C.E.–220 C.E.	Han
220–280	Three Kingdoms
265–589	Period of Disunion
581–618	Sui
618–907	Tang
960–1279	Song
1279–1368	Yuan
1368–1644	Ming
1644–1911	Qing

players, Chinese emperors and their officials presided over an often fractured land of remarkable ecological diversity. However, often with admirable intentions, they gradually, and sometimes catastrophically, weakened this diversity by tolerating widespread deforestation, commissioning often destructive, transitory waterworks, and reclaiming wetlands. No other society in history presided over so much territory so heavily modified by human activity.

Nothing, however, rivaled the hydrological challenges that faced the Chinese: irregular rainfall, frequent droughts, sudden, catastrophic floods, temperature extremes, and unstable river channels, to list only a few. Many

Figure 12.1 *Map of locations in chapter 12.*

rulers took a personal interest in flood-control works and irrigation. They hatched ambitious plans with assiduous care, but their only technology was that of thousands of human hands, armies of workers, who dug tunnels, erected dikes, and dredged waterways. All such works required maintenance every year. Even a short period of neglect caused by a brief military campaign or an unexpected epidemic led almost at once to erosion, breaching of dikes, or heavy silt buildup that blocked key channels. These factors aside, if the state withheld funding, neglect and deterioration soon followed. Without intensive maintenance and continual expenditure, catastrophic ecological degradation was inevitable and frequently occurred.

Over the centuries, water management unfolded in cycles. A strong ruling dynasty with ample funds would commission new waterways and drain lowlands. Productivity would return, until epidemics, war, or other events disrupted maintenance and everything fell apart once more until the cycle began anew. Hampered by simple technology, completely dependent on manual labor, and at the mercy of byzantine webs of officialdom, Chinese water management eventually reached a point of diminishing returns.

THE HUANG RIVER flows through northern China from the Qinghai-Tibetan Plateau, passing through Inner Mongolia, then through the North China Plain to the Yellow Sea.[3] Its yellow-brown silt comes from massive deposits of glacial dust known to geologists as loess, which fills the Huang Basin with fertile soils. Each spring and early summer, silt-laden floodwaters swirl down the river. Over thousands of years, the loess banks on either side of the river formed natural dikes that directed the water across the basin, often at higher levels than the surrounding landscape. The natural dikes often broke, allowing floodwaters to inundate vast tracts of farmland. Disease, famine, and thousands of deaths often resulted. Eventually, the river created a new course for itself, sometimes altering direction dramatically as it oscillated across the floodplain, especially in its lower reaches close to the Bo Hai, the deep indentation where it empties into the Yellow Sea. We would know almost nothing of these long-forgotten disasters but for the recent discovery of a northern village overwhelmed without warning by a breaking levee two thousand years ago.

Sanyangzhuang, Huang River floodplain, northern China: 11 C.E. The daily routine is well under way. Many of the villagers labor in their fields near the close-set compounds of their prosperous settlement. Suddenly, a shout from the fields. A man points toward the high river levee some distance away. A violent flood from rainfall far upstream has breached the earthen bank. Yellow-brown floodwater cascades through a widening gap in the embankment with a dull roar. The levee crumbles; the gap erodes in the face of the rising torrent. The flood roils and tumbles over the gently undulating landscape of fields and villages. For a few moments, the farmers stand motionless. Then they run frantically for their compounds, harnessing their horses to wooden carts. Women grab their young children, leaving cooking fires smoldering, pans where they lie. Grindstones, iron axes, scythes, and plows lie abandoned to the water. The last fugitives load their carts and paddle through the rising flood, lashing their beasts to avoid being bogged down in the shallow water and soft mud underfoot. Within a few hours, the abandoned compounds have vanished. Months later, the receding flood leaves a featureless, uninhabited landscape in its train, with only a few dying trees poking through the fine silt to remind the onlooker that this was once an inhabited place. Centuries are to pass before a new village rises at Sanyangzhuang . . .

This two-thousand-year-old disaster occurred on the gently rolling Central Plain, north and west of the Huang River in Henan Province. In 11 C.E., about a half century before Roman Pompeii vanished under volcanic ash, a catastrophic flood swept over Sanyangzhuang, apparently not the first. In about 2 or 3 C.E., a contemporary chronicler wrote, "Day after day and month after month, it [flooding] became worse. The old places of the water gates were all in the middle of the river. Vast waves overflowed widely."[4] The main channel vanished under the raging waters. History is silent on the subsequent 11 C.E. floods, except for a comment that locusts descended on the land in that year and that the river broke its banks, "overflowing several commanderies." But archaeology and geology have now provided dramatic confirmation of the disaster.

The site came to light in 2003 under at least twenty-seven inches (seventy centimeters) of river silt during the digging of an irrigation canal. So far, four compounds have been excavated, as well as some of the well-tended fields associated with them and a segment of Han Dynasty–era road. Careful borings have revealed traces of at least ten others, some of them dating to earlier times. All four excavated compounds belong to the later part of the Western Han Dynasty and the subsequent interregnum, between around 140 B.C.E. and 11 C.E. Two complete compounds, each surrounded by a wall of densely rammed earth, are now fully excavated. One compound wall has an exterior ditch on two sides. The front gates of each face south, with a well on the south side of the compound and a latrine on the north. The inhabitants dwelled in tile-roofed houses amid open courtyards. Apart from the usual millet and other cereals, the excavators revealed the remains of mulberry trees, used in sericulture (silk growing). The inhabitants had horses for draft purposes, for imprints of their hooves and cart wheels survive in the soft ground surface, as well as those of human feet.

Sanyangzhuang was a prosperous, well-ordered settlement with substantial dwellings that boasted brick floors and foundations of the same material. Each had two or three rooms, with the cooking structure separated from the others. Artifacts of all kinds were found scattered in the rooms and in the open air—grindstones, stone containers, and an abundance of iron tools such as scythes, knives, axes, and iron-tipped plows. In one of the compounds, a pile of fired tiles stood in a tidy stack as if awaiting use to repair or replace a nearby roof. Clearly, the inhabitants had no warning of an impending flood.

Just why the flood occurred remains a mystery. There may have been a variety of causes. The later part of the Western Han period was a time of higher rainfall as well as irregular monsoon events, with frequent droughts. Climate conditions may have changed frequently, with decadal and longer fluctuations in the strength and position of the east Asian monsoon. Human factors may have also contributed to the disaster, among them population growth in Han times that led to migration into the loess plateau regions of the middle reaches of the Huang Valley. At the same time, a rapidly expanding use of iron tools, commonplace at Sanyangzhuang, resulted in agriculture being expanded into hitherto-uncultivated areas, which led to greater deforestation to obtain charcoal for iron smelting, increased land disturbance, and the deposition of higher silt levels in the already silt-clogged river. Greater efforts to canalize and control the river for irrigation with levees and dikes, with usually unsuccessful results, may have also compounded the effects of even minor environmental disasters. The Sanyangzhuang excavations give us a portrait of catastrophic events that occurred again and again over the centuries.

Destructive floodwaters and searing droughts were inescapable here. The earliest hydrological efforts apparently involved the construction of levees for flood control, many of them along tributaries of the main river rather than attacking the Huang itself. Such earthworks kept flood-waters off the lands of those who constructed them. Inevitably, too, in what were troubled times, they soon became instruments of war. One state would divert floodwaters from its neighbor or wash away its grow-ing crops in a patchwork of aggression. By the fifth century B.C.E., levees and dikes were as much fortifications as they were flood-control works. What had once been relatively small-scale local projects mushroomed into much larger enterprises involving the labor of hundreds, even thou-sands, of villagers. A tradition of hydrological works, especially canal digging, on a grand scale began as early as 560 B.C.E.[5]

Long-distance canal building was under way by the fifth century. The Hung Kou canal, built by the kingdom of Wei in 362 B.C.E., drained water from the Huang into a small lake, then back into the river further downstream. Eventually, the canal extended through six small kingdoms and four major rivers to the southeast, taking almost twenty years to build. It was still in use two centuries later. In 246 B.C.E., the Qin state built the Chang-kuo canal, which diverted water from the Ching River to irrigate a huge area on the left bank of the Wei River that extended as far as the northern Luo River. More than 200,000 acres (80,937 hec-tares) came under irrigation, resulting in a dramatic rise in agricultural productivity. The huge food surpluses and wealth generated from this ambitious scheme helped fuel Qin's conquest of its rivals, which culmi-nated in the unification of China under Emperor Qin Shi Huang in 221 B.C.E. The Qin Dynasty soon faltered, as did its canal, which silted up badly within a century and a half.

By no means were all Qin projects failures. In 256 B.C.E., people liv-ing along the banks of the Min River, in Sichuan Province in western China, on the southern frontier of the Qin domains, suffered greatly from snowmelt flooding. The Qin governor, Li Bing, rejected the idea of a dam, for he also had the responsibility of keeping the river channel open for military vessels. Instead, he built the artificial Dujiangyan levee that redi-rected part of the river flow, cutting a tunnel through nearby Mount Yulei to discharge the excess water onto the dry Chengu Plain.[6] Thousands of people labored for four years on the levee, constructed of long, sausage-shaped baskets of woven bamboo filled with stones and held in place by wooden tripods. Li Bing used fire and water to heat, then cool the hard

Figure 12.2 *Dujiangyan. A young visitor examines the wooden tripods and woven bamboo baskets filled with stones that form part of the irrigation system on the Min River. (Bobby Yip/Corbis)*

rock of the sixty-five-foot- (twenty-meter-) wide tunnel, and the task of carving it took eight years. The project proved a major success, eliminating flooding and creating irrigation works that have fed large numbers of people ever since. It still irrigates over 2,046 square miles (5,300 square kilometers) of farmland and survived the Sichuan earthquake of 2008 with only some cracks in the levee.

The Han emperors, who presided over China from 206 B.C.E. to 220 C.E., established patterns of imperial government and water management that were to persist, albeit with constant modifications, right into the early twentieth century. They undertook at least forty major water projects on the Huang River and its tributaries alone. Much of the work involved not irrigation but flood control—building levees over long distances and even turning the river in different directions to regulate its flow. The Huang itself required much larger earthworks than its tributaries, to the point that the river now flowed at a higher level than the houses and fields on either side, thereby increasing the risk of catastrophic flooding in the event of a breach. In undertaking these backbreaking tasks, the Han set precedents for water management that endured for centuries.

Chinese practice differed from that in the West. The Han and their successors relied on large numbers of workers to build dikes, using but the simplest of tools, a practice that remained unchanged for two thousand years. As early as the second century B.C.E., there are references to "several tens of thousands" of people laboring on projects that lasted two or three years. By the seventh century C.E., a staggering five million villagers are said to have worked on one canal alone over twenty years. Just what social and political mechanisms organized these armies of workers is much debated, but the unification of China allowed the emperors to mobilize much larger numbers of people than hitherto.

Water management achieved military and social goals. The rulers of the Sui Dynasty (581–618 C.E.) started construction of an ambitious canal system that linked Hangzhou, in the southeast, to the southern capital of Yangzhou, on the Yangtze River, then ran northward to the Huang River and beyond. Mongol emperor Khubilai Khan extended the Grand Canal northward in the late thirteenth century, linking the Huang and Beijing, for a total length of over 1,100 miles (1,770 kilometers).[7] Over 2.5 million workers labored on the extension of the canal and the paved highway alongside it. This troubled waterway facilitated grain transport and trade between different ecological zones of the state—provided it was properly maintained, which it often was not.

At one level, waterworks of all kinds, especially levees and irrigation works, were the responsibility of individual communities and groups of villages. At another, the state was involved, for it had both the overarching authority and the funds to undertake much larger projects beyond the means of humble farmers. The government organized corvée labor for major projects. Sometimes the authorities compelled people to work. At other times, they used the practical, longer-term benefits of the project to urge participation. Many workers were even paid in rations to labor on flood works. Much depended on the relationship between the authorities and the local people, and on the probity and commitment of the provincial officials who carried out the emperor's orders.

The Han consolidated water management under a central office, the Director of Water Conservancy, part of the Ministry of Public Works.

Figure 12.3 *The Grand Canal at Suzhou. (Dean Conger/Corbis)*

This was a planning organization, with responsibility for recruiting labor and construction left in the hands of local officials. The key figures
were the water managers, who got their hands dirty in the field. It was
they who repaired levees and bridges, cleared irrigation canals, drained
floodwaters, and dealt with dike breaches and other disasters. Some of
them achieved lasting fame, among them Wang Ching, who is said to
have stabilized the main channel of the Huang River so effectively during the first century C.E. that there were no major dike breaches for one
thousand years. He dredged key channels, strengthened key levee points,
installed hundreds of sluice gates, and dug new channels for tributaries.

The Huang never relaxed, changing course abruptly, delivering devastating floods, and defeating all manner of ingenious strategies. Should
one, for example, confine the river channel between high levees up to 1.9
miles (3 kilometers) apart, or allow it to flow over a front of as much as
6 miles (10 kilometers) between lower embankments? A narrow channel
caused rapid erosion of the riverbanks. A wider channel brought rapid
silt accumulation and could also kill the farmers who moved onto the
fertile, damp channel land to cultivate it. Conflicting philosophical and

religious beliefs complicated the debate. None other than Yu the Great had established the ancient precept of "using the river to control the river."[8] Taoism, with its deep roots in the remote past, indeed urged that nature be allowed to do its work. But the more recent, contradictory doctrines of Confucianism held that humans should work to control nature.

The state's efforts at water management were called *shuili*, or "water benefits." River works were part of *shuili*, the building of dikes or levees, sluice gates, weirs, and dams, as well as canals. This involved expensive, large-scale organization, with the cost of such works usually falling to the central government. The political equations involved were complex— balancing the common good against the competing interests of communities upstream and downstream, and protecting the common good from the ambitions of unscrupulous landowners. *Shuili* in the narrower sense involved irrigation works regulating water for agriculture, more a major concern in rice-growing areas in the south, for irrigation was not that common in the north, even if some emperors tried to introduce southern techniques there. These included drainage and land reclamation, the diversion of waterways, digging irrigation canals, and creating rice fields. All of these activities were of importance in a state where a high proportion of the subjects dwelled near complex river systems. No dynasty took these responsibilities more seriously than the Qing.

THE QING, NON-CHINESE Manchus from Manchuria, deliberately adopted Chinese ways when they seized power in 1644. Qing rulers created a highly authoritarian and centralized domain, bolstered by strict neo-Confucian moral orthodoxy. For the next century and a half, they presided over a "Golden Age" of prosperity and sound government. The emperors of these prosperous times paid careful attention to the rivers near Beijing, partly because they lay close to the Grand Canal. Almost invariably, a major flood event early in an imperial reign would trigger a new river-management campaign. The emperor would show symbolic concern for the people, galvanize his bureaucracy, and spend money on major works. Inevitably, however, inertia would set in, until more flooding started the entire process anew.[9]

During the mid- to late seventeenth century, life in northern China was harsh and tumultuous, following an era of warfare. Climatic conditions were unsettled. The major problem was prolonged droughts, like that of 1691–92, which required massive relief efforts, using the Grand Canal from the south. The loss of life was remarkably small, thanks to the state's proactive response, driven by strict imperial supervision. The rulers took a particularly close interest in the rivers near Beijing, which suffered from the Huang River's tumults and also affected the navigability of the Grand Canal. Emperor Kangxi (1661–1722) personally supervised dike construction along nearby rivers, inspecting flooded areas and starving communities. In 1698, he ordered that the Yongding River, the largest and most chronically unpredictable waterway flowing through the capital, be dredged and that a long protective dike be constructed. Completed in four months, the earthwork collapsed in many places during the next year's floods. The emperor measured silt accumulation in person, ordering new levees that drained water into nearby marshes. This time, his stabilization efforts worked. Kangxi knew that his personal interest had tipped the scale. "As for water that is not clear, I have a policy to manage it," he proclaimed triumphantly. Officials hastened to carry out his orders. He also created a river-patrol organization in 1698, which supervised two thousand soldiers as they monitored the Yongding's riverbanks. The river became known as the "River of Eternal Stability."

Kangxi was an exemplary ruler, and his son Yongzheng (1723–35) was just as efficient. When heavy summer rains inundated seventy-four counties and the land became a "vast ocean," he remitted all taxes owed by victims and commissioned an ambitious water-management network supervised by his brother, Prince Yi. In four years, Yi organized the digging of canals and ditches, as well as extensive tracts of rice fields, especially in low-lying areas west and south of the capital. The local people preferred to eat millet and wheat, so the government purchased their rice so they could buy grain. After Yi's death in 1730, rice cultivation declined, but the effects of his management endured. Administration of the Yongding River was almost military, with its numerous supervisory officials and well-oiled routine. All dredging was completed before laborers had to harvest wheat and before the rain that fell on either side of the harvest. The big floods arrived in late summer, when the local governor or his deputy moved to the river to supervise emergency repair work.

Crisis after crisis descended on the land. In 1738, a month's rain destroyed crops over a wide area and inundated the homes of over twenty thousand people. The disaster affected over 1.5 million Chinese, spurring fears that the wet soil would inhibit the planting of winter wheat, leaving people hungry a year later. A catastrophic drought in 1743–44 led to two to three years of expensive river works, but no one, least of all the emperor, had any illusions about the long-term outcome. Fortunately, with a full treasury and widespread prosperity, the state could spend money on effective famine relief, especially on rice shipping along a properly maintained Grand Canal.

The effectiveness of waterworks throughout China depended on strong, proactive administration, both centrally and locally. A water manager didn't only have to be an excellent engineer. The techniques for canal building and dike maintenance were familiar to all. He also had to be adept at handling people involved in the planning and the actual work, as well as the complex logistics needed to appropriate raw materials for the job at hand. Above all, he had to realize that local governments were the most vitally concerned in successful water control. Managers with these qualities were rare, but among them was the Qing official Chen Hongmou, a master of hydrological administration.

CHEN HONGMOU (1696–1771) served as governor-general and in lesser posts in more than a dozen provinces scattered across the empire between 1733 and 1763.[10] Like the Roman bureaucrat Vitruvius, he was devoted to public service in the highest sense. This was a consummate, trustworthy bureaucrat with all the best qualities of the breed. No one could call Chen an original thinker, but he stood out for his energy and thorough attention to local problems. He was, above all, revered for his statecraft, *jingshi*, or "ordering the world." He was a stern, moral believer in benevolent government, who also strongly supported leaving responsibility in local hands. And nowhere did he display these qualities better than in his administration of economic development—regarding land, the people, and *shuili*. On several occasions, he was selected for a provincial post for his expertise at water management.

This remarkable official developed a well-oiled administrative machine for his ambitious hydrological projects. He instructed his subordinates to pay careful attention to local terrain and cropping patterns, to make the most profitable use of the local environment. Most important of all, the frugal Chen encouraged very expensive long-term investment as the best way to achieve success, rather than short-term penny-pinching. In Yunnan Province, in southwest China, where he worked in the 1730s, Chen inherited a province with little irrigation agriculture and a growing population. He drew up a comprehensive plan for his superiors that aimed at making the province agriculturally self-sufficient. Yunnan is mountainous, and the only source of irrigation water was runoff. It was easy enough to channel the water, but much harder to distribute it over wide areas. Chen devised a series of mountain reservoirs, each regulated by several dams to collect rainwater. He ordered Yunnan's magistrates to survey the terrain and water flow of their districts, and he himself compiled the information into a hydraulic atlas. He also instructed them to assess the ability of individual landowners to pay for the improvements, crafting elaborate checks and balances to minimize corruption. Then he divided the project into sectors, creating a complex network of assessments, government loans, and direct official financing from rental income to pay for the work. A liberal tax policy gave landowners incentives to participate. In the end, a network of tile gutters and bamboo pipes irrigated wide tracts of fields, using gravity reinforced by strategically placed waterwheels.

The scale of Chen's projects seems almost incredible, until one realizes that he had almost unlimited labor at his disposal. Between 1747 and 1751, he dug more than 32,800 irrigation wells in eight counties in Shaanxi Province, an arid landscape west of the Huang River. The benefits were immediate: Some cultivation was possible even in drought years. The growing of vegetables skyrocketed. Much depended on his energy. When Chen was absent, the pace of well digging slowed. In late 1751, he started a project of land reclamation and digging of irrigation wells, financed by government loans to aspiring landowners. By this time, though, the limits of well digging were apparent. So instead, he planned a network of irrigation canals that drew water off mountain streams. He molded public opinion by calling town meetings, many of which he attended in person. Villages were required to provide labor; if the requirement was too onerous, Chen provided grain to hire workers. As was customary, he expected local magistrates to pick up the tab for materials, pointing out that the increase in productivity would more than justify the initial expense. He was correct. When the topography militated against gravity, he installed waterwheels of a design widely used in southern China and still virtually unknown in the north.

In 1751 and 1752, Chen tackled a major drainage problem in the northeastern part of Henan Province, where Huang tributaries regularly inundated much of the Gui'de prefecture, blocking the single natural drainage channel with silt. Famine followed even moderate harvests. He devised a massive scheme to dredge all the waterways in the region. Chen believed that the state should never proceed arbitrarily with any form of hydraulic project without consulting the people—the stakeholders, in modern parlance. He also thought that the state should alert the stakeholders to the potential of new work, the official's moral task being to "civilize and instruct." In the case of Henan, he assigned the responsibility for much of the work to village leaders and landholders, insofar as this was practicable, while investing enormous sums of government money to carry out work that local agencies could not afford. Perhaps most important of all, he enlisted the assistance of authorities in the neighboring province of Jiangsu, using his own personal connections. Contemporary accounts record a dramatic agricultural recovery that endured for generations.

Chen argued that extensive water-management projects were in the general interest. He was proactive in repairing existing facilities, consulting archives to check on how they operated, and did not hesitate to commit official funds when an entire system required configuration. In managing the work, he did everything he could to foster a sense of common purpose and to encourage teamwork, which required placing considerable responsibility on individual shoulders. He disliked any form of forced labor, on the grounds that it was less productive than hiring workers for rations or using employment on hydrological projects as work relief. The strategy, said Chen, employed non-farmers, reduced administrative corruption, and improved the infrastructure all at the same time.

CHEN HONGMOU WAS a thoughtful administrator, but schooled in a tradition of massive water projects involving thousands of people. He was so efficient that the emperor only allowed him to retire when he was terminally ill. By the end of his life, there was a mounting anxiety about the mushrooming imbalance between growing farming populations and agricultural production. There appeared to be no "natural" way of taming rivers to serve farmers. Gradually, Chen's successors, living in more troubled times and working with inadequate funds, realized that the major hydrological achievements of the past, involving thousands of laborers, could never be repeated or maintained, nor would they ever achieve the levels of productivity that theoretically lay ahead.

A major flood in 1761–62 again showed the ineffectiveness of river management in the north. Then came the flood of 1801, early in the reign of the emperor Jiaqing (1796–1820), when floodwaters from both rainfall and mountain runoff inundated parts of the imperial palace. Heroic and expensive flood-relief efforts involving at least fifty thousand workers, personally supervised by the emperor, ensued, with the court keeping a close watch on local river officials for any signs of malfeasance. These pricey efforts were of little avail. Flood after flood caused widespread damage and suffering, as well as a persistent and serious drain on government coffers. Centuries of channeling and diking led to massive ecological disasters year after year, when the landscape resembled a lake, or when successive droughts parched crops and thousands perished. There was no way to reverse the long-term environmental decline. The labor needed to maintain even the existing irrigation works was unsustainable.

THE FAMINE-RELIEF policies of the Qing rulers focused on an empire-wide granary system, which collected surplus grain after the harvest, then sold it during shortages to keep prices stable.[11] In times of severe hunger, the state shipped grain from different areas to provide meaningful relief to famine-plagued regions. The Qing also embarked on long-term measures to increase agricultural productivity, including tax incentives, resettlement of migrants, promotion of new cropping methods, and new waterworks. Rice in particular became an important commercial crop in the breadbasket of the south. Chinese hydrologists, like their preindustrial counterparts elsewhere, had to rely on manual labor to achieve their goals. Their technology extended little beyond hand tools, although they developed, or used, a wide variety of milling and pumping devices, many of which were used for rice farming.[12] Many farmers used versions of the *shaduf*, which arrived in China as early as the fifth century B.C.E. People lifted well water with simple pulley mechanisms; hand-operated paddle wheels swept water into flumes. Chinese farmers invented the square-pallet chain pump, an endless chain of pallets that pulled water upward through a trough into the field below. Some were turned by hand, but more commonly they were turned by foot power or by an ox. In the first century C.E., watermen working day and night used chain pumps to hoist water from the Lo River, by Luoyang in northern China, for the city's water-supply pipes. The chain pump spread widely into the West, there being a fine example at Hampton Court Palace, near London, that was installed to remove sewage in about 1700 or perhaps earlier. Waterwheels were in widespread use in later times, used not only for irrigation but also for city water supplies and, at Louyang, to fill a large public bath, said to be able to hold thousands of people. Chinese milling technology included trip-hammers, horizontal grain mills, and other devices, including metallurgical bellows operated by waterpower. None of these simple but often ingenious technologies proved of use when confronting river floods.

Henan lies at the southern end of China's temperate zone, an area about the size of the United Kingdom, surrounded by mountains but open in the north to the Yangtze floodplain. Until the eighteenth century, Henan had a sparse, war-devastated farming population, which subsisted off rice cultivation. Most cultivators lived in river valleys and on the lowlands around Dongting Lake, where rich alluvial soils washed down from the mountains provided excellent farmland, some of the best in China for rice farmers.[13] They relied on irregular monsoon rains, which coincided with the annual Yangtze flood. Most years, water backed

Figure 12.4 *Chain pump. (After Joseph Needham,* Science and Civilization in China, *vol. 7, p. 309. Courtesy: Cambridge University Press)*

up from the river into Dongting Lake, blocked the Henan basin drainage, and caused widespread flooding.

Today, Dongting Lake is the second-largest lake in China, with an area of about 1,042 square miles (2,700 square kilometers), but it has a long and complex hydrological history. Enormous amounts of water flow into the lake from the Han and Yangtze rivers and other waterways to the south. The gradient of the lowlands is so slight that water flows very slowly, the lake level rising and falling dramatically during the year, with the highest level coinciding with the Yangtze flood in August and September.

Figure 12.5 *High-lift* norias *photographed by Joseph Needham in the 1940s. (From his* Science and Civilization in China, *p. 312. Courtesy: Cambridge University Press)*

Anticipating an influx of migrants during the eighteenth century, the Qing intervened with funds to invest in major dike works to convert swamps into rice paddies. At first, things went well. Up by the Yangtze, farmers enjoyed the protection of long, closely supervised dikes. Settlers lived behind these enclosure dikes, often farming below lake level. Local families controlled these units; there was much less government authority here. When the floods came, the water rose slowly enough for people to move away. Under minimal government supervision, wealthier landowners took advantage of this to extend their enclosures progressively further into the lake, gradually reducing its size. A heavy silt load from the Yangtze also contributed to the shrinkage.

At first, the state enjoyed large tax surpluses and a stable political climate, so it could afford massive investments, stimulated in part by a disastrous flood in 1727 that caused 430 dike breaks. Tax incentives also encouraged private dike building, but conflicts soon arose between those who wanted to clear more land and those who did not. Wrote Chen Hongmou in 1757, "Dikes on the lake fill the view endlessly like fish scales and there are cases of fighting with the water for land."[14] Eighty years later, the lake was a maze of illegal dikes and becoming shallower and shallower as the population continued to rise. Disaster was inevitable.

A major flood in 1831 caused widespread suffering, followed by frequent, exceptional inundations from 1834 to 1879. The 1881 government gazetteer reported that "not one single dike remains here, neither official, people's, nor illegal."[15]

Dongting Lake dramatized the long-term tensions between the state and local interests in China. The state was neither an all-controlling despot nor in the habit of leaving the countryside alone. Rather, the imperial bureaucracy intervened actively when the local rural economy demanded it, especially by supporting major waterworks. When the private sector assumed greater importance, as it did with rising populations and increased commercialization of agriculture, conflicts between government and local interests intensified. By 1800, a combination of natural disasters, defiance of imperial edicts, and rising local conflicts had set the stage for much larger disorders that unfolded during the nineteenth century.

FARMERS IN BOTH northern and southern China are at the mercy of climatic forces far from their homeland, which aggravate the perilous circumstances of subsistence agriculture, especially in the north. El Niños, the most powerful drivers of global climate after the passage of the seasons, influence the northerly movement and intensity of monsoon rains. Sclerotic administration, political turmoil, widespread corruption, and inertia caused as much havoc in the north as they did in the south.[16] During the 1850s, millions of farmers fled from their homes in the face of massive flooding, into the arms of rebel armies engaged in the bloodiest civil wars of the nineteenth century. Then, in 1876–78, the summer monsoon faltered when a major El Niño developed in the southwestern Pacific. Northern China baked under a pitiless sun. Summer and autumn crops withered. Throngs of emaciated refugees descended on towns and cities. In Shanxi Province, there had been drought in 1875. The renewed drought brought catastrophe. Hungry peasants ate their seed, then the husks, turnip stalks, and wild grass. They sold the timber of their houses and devoured the rotten reeds from the roofs. Finally, they sold their children or consumed human flesh. No one knows how many millions perished in a disaster compounded by a complete breakdown of relief shipments. The overgrown Grand Canal was impassable for hundreds of miles. Silt from the Huang River flowed through breached dikes and filled a canal that had once been the pride of Qin emperors. Diplomats and missionaries reported horrifying scenes. "In the ruined houses the dead, the dying, and the living were found huddled together . . . and the domestic dogs, driven by hunger to feast on the corpses everywhere to be found, were eagerly caught and devoured."[17] Even when the monsoon returned in 1878, recovery was difficult, for seven tenths of the population was dead. Only a third of the normal grain was sown. Those who harvested it had to guard their crops against desperate neighbors or suffered from dysentery from unaccustomed food. It took at least two years for northern China to recover.

Like other preindustrial civilizations, China had reached beyond the limits of sustainability at a time of unusually severe El Niño activity, which persisted into the early twentieth century. Thousands of years of regimented labor had failed to master the inexorable cycles of the natural world.

PART IV

Ancient American Hydrologists

Lords, rituals, and gravity: In which we return to the Americas and enter the complex water world of the ancient Maya, where kingship and water ritual provided a unique framework for water management that ultimately faltered, then failed. In contrast, the Andeans displayed a brilliant mastery of respectful hydrology in deserts, rainforests, and mountainous terrain alike.

CHAPTER 13

The Water Lily Lords

"AND WHEN WE SAW all those towns and villages built in the water . . . and that straight and level causeway leading to Mexico, we were astounded. These great towns . . . and buildings rising from the water all made of stone, seemed like an enchanted vision." Spanish conquistador Bernal Díaz del Castillo never forgot that memorable day in November 1519 when Hernán Cortés and his small band of adventurers gazed down from the mountains on the Aztec capital, Tenochtitlán, the "Place of the Fruit of the Prickly Pear Cactus." Tenochtitlán dazzled the Spaniards. At least two hundred thousand people lived in or around the city. Five or more major canals flowed through it, with secondary channels forming rectangles that dissected the metropolis into large blocks. Four causeways divided Tenochtitlán into quarters, intersecting at the walled central plaza, with its spectacular temples. The entire central precinct blazed with bright colors, a pageant of serpent-studded walls, feather banners, and brilliantly plumed capes and uniforms. The pyramids of the sun god Huitzilopochtli and other deities rose like small mountains around the central plaza, giving a profound impression of stability, majesty, and overwhelming power. And overwhelming power it was. At the Spanish Conquest, the Aztecs presided over a tribute empire of more than five million people. "Today all that I then saw is overthrown and destroyed . . . nothing is left standing," wrote Bernal Díaz decades after the conquest.[1] Ancient America's greatest city now lies under the urban sprawl of Mexico City.

Over thousands of years, Native Americans developed a remarkable expertise at both agriculture and water management in every kind of

Figure 13.1 *Map showing locations mentioned in chapter 13.*

environment imaginable, from deserts to tropical rainforests, in the low-lands and the highlands of Mesoamerica. Tenochtitlán's rulers drew on a huge reservoir of hydrological expertise. To feed the city's enormous population meant using every acre of the landscape, using fertilizers, growing cacti on thin-soiled hillsides, and importing tropical crops from the lowlands to the capital's enormous market. Aztec farmers used springs and furrow channels, harvested mountain runoff, built hillside terraces, and relied on floodwater irrigation. Above all, however, adequate food supplies depended on efficient use of wetlands. Tenochtitlán merged into the marshes of the lake that once filled much of the Valley of Mexico. Two shallow basin lakes, Chalco and Xochimilco, were covered with artificial, islandlike gardens known as *chinampas*. By laying out a vast network of drainage ditches, the farmers created these gardens, at first at the edges of islands and peninsulas, then rapidly further out across the lake. In time, the *chinampas* became regularly laid-out and carefully planned wetland plots. Each consisted of alternate layers of lake-floor mud and vegetation, protected by "fences" of trees to retain the silt. The dark, organic soils were highly productive—watered with large pots or by splashing water from canoes while paddling along the waterways that intersected the fields. The farmers simply loaded their harvests into their canoes and paddled them to the market in the capital.

The labor required to establish a *chinampa* system was enormous and really only economical when dry agriculture could no longer feed growing populations. Once reclaimed, wetland gardens had fertile soils and a high water table that protected growing crops against frosts, an important consideration in the Valley of Mexico. The Aztec state reclaimed thousands of acres of such wetland, settling farmers on small rectangular plots where they grew crops and the all-important flowers for the temples.

Chinampas were nothing new in the Valley of Mexico, but their great expansion came during the fifteenth century as Tenochtitlán mushroomed into a huge city. A thousand years earlier, another huge city, Teotihuacán, in the northeast of the Valley of Mexico, fed a teeming urban population off extensive tracts of swamp gardens, but little is known of its agricultural practices. There is no question, however, that the wetland farming practiced by the Aztecs had an ancestry deep in the past, dating back to when Teotihuacán was but a village, in the first millennium B.C.E. Wetlands and agriculture went together in other areas of Central America, too, among the fabled Olmec people of the Veracruz region, one of the founding societies of Mesoamerican civilization, and also among the ancient Maya.

RICHLY CAPARISONED LORDS, intricate glyphs, great temples hidden in dense rainforest—the exotic Maya civilization of lowland Mesoamerica has captivated us ever since travelers John Lloyd Stephens and Frederick Catherwood revealed its flamboyant temples and art to an astonished world in the 1840s. The fiendishly complex Maya script baffled generations of epigraphers until its successful decipherment over the past forty years, one of the greatest scientific triumphs of the twentieth century. Originally, everyone thought of the Maya as peaceful astronomer-priests with an obsession with calendars and the stars. Decipherment has painted an entirely different portrait, of a perilous, ever-shifting political landscape. Great lord competed against great lord; volatile polities vied with one another in seesaws of diplomacy and short-lived military campaigns. Maya civilization was never a unified state like that of the Egyptians. The rainforest of the lowlands supported a jigsaw puzzle of ceremonial centers large and small, whose fortunes ebbed and flowed over many centuries.

Maya civilization developed in the lowlands of what is now Guatemala, Honduras, and Mexico during the first millennium B.C.E. Its roots lay in much earlier subsistence-farming communities, some of which grew to prominence, each with its own deities and temples. Many Maya beliefs and institutions came from the still little-known Olmec societies of the Veracruz region, whose leaders traded with highland peoples in the interior before 1000 B.C.E. Maya civilization developed well before 600 B.C.E. Before the time of Christ, great ceremonial centers flourished at El Mirador and other locations. In the centuries that followed, Maya populations exploded as new centers like Tikal and Naranjo came to power in the southern lowlands of what is now Guatemala's Petén region. Subsequent Maya history is a kaleidoscope of rising and falling kingdoms and volatile alliances, rivalries, and implosions.

The brilliant success of this remarkable civilization resulted in considerable part from sophisticated water management in a unique environment without great rivers or heavy and relatively dependable monsoon rains.[2] The Maya lived in highly varied tropical environments, a patchwork of coastal plains, scrub, and dense forest. Their manipulation of water and land was so successful that more than one hundred centers large and small and thousands of villages flourished between 695 and 900 C.E. For all its political volatility, Maya civilization expanded successfully until the ninth century, when it suffered a major collapse in the southern lowlands in the face of drought, environmental degradation, and political unrest. Cities vanished into the forest; population densities tumbled as the survivors returned to village life. Elsewhere, especially in the northern Yucatán, where there was subsurface water with relatively easy access, major ceremonial centers and powerful lords thrived until the Spanish Conquest of the sixteenth century. Unraveling the complex history of Maya civilization has barely begun, but it's clear that water and water management play major roles in the story of rise and collapse.

The southern Maya lowlands covered much of Belize, central and northern Guatemala, and neighboring areas of Mexico and Honduras.[3] The Maya depended entirely on unpredictable subtropical rainfall. Dense forest once covered the landscape, occasionally giving way to patches of open savanna covered with coarse grass and stunted trees. Today, at first glance, the still-dense tree cover seems monotonous, but the seeming uniformity hides an astonishing diversity of local habitats, all of which presented special challenges to ancient Maya farmers. This is a hot, humid, and unforgiving world, where the soils are generally of only moderate fertility, except in parts of the Petén and along larger river valleys. The best soils lie on the hills and on better-drained flatlands. Much of the lowlands is poorly drained swampland, soggy for some of the year, hard and dry for the rest, and thus almost impossible to cultivate. To take one area alone, no less than about 40 percent of the most densely settled zone of the Maya lowlands, between Tikal and Río Azul in the Petén, comprises such swamplands, commonly called *bajos*, only part of which are of economic use.

Unpredictable, seasonal rainfall and a dry season of four months or so between January and April compounded the Maya farmers' difficulties. There are few permanent rivers or streams here to ease the water shortages of the dry season. Where there are such sources, the flow moves slowly, poor drainage conditions allow little absorption, and water levels fluctuate only slightly during the year. Classic forms of irrigation based on gravity do not work here. Adding to the farmers' water problems, there are few natural springs. Especially in the northern Yucatán, the Maya built wells to access groundwater, often near natural depressions. Northern Maya farmers also made use of cenotes, water-filled natural sinkholes that they sometimes modified to form walk-in wells. Generally, however, the Maya lived with constant water deficits, and their society was extremely vulnerable to prolonged droughts. It's a tribute to the ingenuity and opportunism of ancient Maya farmers that they managed to support a large number of elaborate ceremonial centers and feed non-farmers for a thousand years. To understand how they did it, you have to look behind the spectacular façades of their great ceremonial centers at the realities of village life.

BY THE THIRD century B.C.E., Maya farming communities flourished on the coastal plains and along permanent rivers and streams where water abounded.[4] As population densities rose, farmers moved away from the rivers and settled along the margins of natural sinks and *bajos*. In their natural forms, such depressions would not support permanent settlement. When the farmers modified them and enhanced their natural water-retaining characteristics by turning them into small reservoirs, the potential of the landscape changed profoundly. The stability of each community depended on clean water from some form of reservoir, following a type of "one village, one tank" principle that endured not only here but also in regions such as southern India and Sri Lanka for many centuries, and does to this day. The larger the reservoir, the larger the village remained—this fundamental reality defined Maya life and civilization even in the days of the great cities and mighty lords.

Standing water quickly becomes stagnant and breeds parasites. At some point, the Maya discovered that water lilies, *Nymphaea ampla*, proclaimed the presence of clean water.[5] These sensitive, hydrophytic plants can grow only in water shallower than 9.8 feet (3 meters) that is relatively free of algae and calcium. Still water without the lilies was undrinkable. It was no coincidence that the water lily eventually became a major symbol of royalty in Maya society, depicted on monuments and polychrome painted pots. Some Maya rulers even became known as Nab Winik Makna, or "Water Lily Lords." The flowers often formed part of lordly headdresses, and many prominent leaders bore names that included "water," such as the Water Lily Lord of Tikal and Lord Water of Caracol.

At least two thousand years ago, the Maya began to symbolize the cosmos in their iconography, a world peopled with deities and powerful

Figure 13.2 *A Maya lord at Copán, Honduras, wears a water lily head-dress. (From Group 10L-2. Courtesy: Barbara Fash)*

supernatural forces, among them the Water Lily Monster, a zoomorphic creature bearing the pads and blossoms of the water lily, who symbolized lakes, rivers, and swamps.[6]

Here, as in Africa and Bali, land, ritual beliefs, and water were completely intertwined, for the issue was not an assumed abundance, but water as an independent variable in human life in ways that would have been unimaginable to, say, an ancient Egyptian. Constant ritual activity surrounded water in Maya life from early times, especially ceremonies involving the ancestors, for here, as in other farming societies, those who had gone before were a great moral force in human existence. They played, and still play, a vital role in ensuring the continued fertility of the land, ample rainfall, and good health. To this day, for the Maya, life constantly ends and begins anew—with a birth, a family death, the abandonment of a house, the building of a new one. The emphasis is on renewal, whether it be a household, a village, or a major temple. When an ancient Maya shrine was ritually destroyed, rulers burned incense and left broken items as offerings, before a new temple was built over the old and eventually dedicated with entirely new objects. Over the centuries, these once-simple rituals of dedication and renewal, of ancestor worship observed in villages large and small, morphed into elaborate ceremonial observances developed by generations of rulers, who transformed the superstructure of Maya society. There are surprising connections between this elaboration and the story of water management in the lowlands.

THE MAYA DWELLED in an environment of constant disruptions: drought and crop failure, torrential rains and soil erosion, unexpected storms that ruined their plantings. They farmed with the simplest of technology, but with a brilliant acumen that took advantage of their diverse forest landscape. Like tropical farmers elsewhere, they used slash-and-burn methods, often called milpa cultivation, cutting down forest, burning the wood and brush, and working the ash and charcoal into the soil. The average plot could produce a crop for two years before being abandoned, for the soil lost its fertility rapidly. The farmers also terraced steep hillsides, the stone-faced terraces trapping silt that would cascade downslope during violent downpours. Like people on the highlands, they also reclaimed extensive tracts of wetland, which they turned into networks of highly productive raised fields, somewhat like the Aztecs' *chinampas*.[7]

The first arrivals settling in new areas founded their villages close to the best water sources, in places where there was abundant farming land and natural sinkholes that collected water during the rainy season. The situation was exactly the same as it was among the Hohokam in the Arizona desert: Those who acquired the best land and water sources acquired power from their control of such resources. As early as 1000 B.C.E., some Maya communities were already manipulating water supplies. By then, farmers in what is now northern Belize were digging shallow ditches to drain the edges of swamps. As people moved away from major rivers, new water systems developed. By 300 B.C.E., the inhabitants of Cerros, in central Belize, were reclaiming wetlands. As they quarried rock for increasingly large buildings, they created both reservoirs and canals that stored water and also drained the central precincts during the rainy months. Through ingenious use of rock sills extending into canals and reservoirs, they conserved water during the dry months.[8]

As time went on, Maya water-storage systems became ever more elaborate. At a much larger community, Edzná, in Campeche, the farmers may have developed a system of linear reservoirs that also served as waterways for heavily laden canoes, but the evidence for this is incomplete. Navigable waterways were at such a premium that the Maya transported virtually everything on human backs, which placed immediate limitations on their farming and irrigation activities. The Maya tended to position their communities at the margins of swamps, where they could maximize the potential of raised fields and gardens that could be drained of floodwaters. Everywhere, they took advantage of large depressions in low-lying terrain, which they modified to store water. Perhaps, too, some communities also worked over internally drained *bajos* so that they maintained predictable water levels for raised fields, easily watered by hand with the water that surrounded them. Some of the earliest Maya centers, like the great city of El Mirador, came together as early as 300 B.C.E. but imploded several centuries later, in part because of the silting of their reservoirs based in natural depressions, what one might call a concave system of water conservation. Later water managers learned from mistakes made at El Mirador and thus avoided similar problems.[9]

DESPITE THE COLLAPSE of El Mirador and other early centers, attributable in part to water problems, the Maya developed a remarkable expertise with water conservation and management both at the village level and for rapidly growing major population centers. Every household, every community, celebrated water rituals, but, inevitably, the first founders to settle away from permanent rivers, lakes, and streams began to distinguish themselves from later, less privileged arrivals, because they controlled the best water supplies and farmed the most fertile land. Inevitably, too, some of them had dreams of grandeur. Some adopted iconographic symbols from Olmec lords, the earliest of all lowland rulers, whose deeds were the stuff of legend. The founders built themselves larger dwellings and constructed small temples. They traded exotic objects from afar such as jade, a symbol of wealth and prestige, and could afford to make offerings to the ancestors of considerable value. Many of these activities required labor from what were now becoming commoners, mere subsistence farmers. The emerging rulers rewarded such labor not only with food and exotic luxuries but also with feasts and increasingly elaborate rituals to which they invited the entire community. Complex issues of competition for labor and prestige were involved, for each founder family, each emerging elite, had to compete with its neighbors with similar ambitions for both commodities. The inconspicuous catalyst for many of these developments was water; the powerful undercurrent was the unspoken power of villages that were basically self-sufficient.

Large reservoirs and imposing public buildings went together, for the one provided the material for the other. By 200 C.E., some of the founding settlers had become powerful rulers, especially if they controlled water sources. Not that these sources were necessarily potable, for the low levels of the dry season caused the water to turn murky and disease ridden. Wealthy rulers could pay for the repair of water channels or field systems damaged during heavy rainstorms, but those services did not come cheap. Each ruler demanded as much in labor and in kind as the marketplace could stand, but they had to be careful, for, in the end, villagers could quietly move away into the forest or into the realm of a ruler with fewer demands. Any form of more centralized Maya water management involved the right topography, highly flexible labor management, and a great deal of trial and error. Above all, it required the loyalty of a dispersed village population to individual lords. These rulers went to enormous lengths to maintain such ties. Their prosperity, prestige, and very existence depended on it.

Once village farmers moved away from permanent water sources in the face of a growing population, they had to rely almost entirely on reservoirs. Water demand rose as the population increased, so artificial reservoirs became larger and larger. One can imagine teams of men laboriously digging into swampy clay and hefting basketfuls of it onto

more-solid ground. They stand knee-deep in clinging mud, sweating pro-
fusely in the still, humid air. Women pile clay near the swamp, where
artisans puddle it with lime to fashion stucco. Other villagers lever large
blocks of limestone from the edges of what will become a reservoir, rolling them on logs to waiting masons, who trim the rock into blocks or set
aside irregular boulders for the rubble core of the pyramid that rises behind them . . .

Plentiful water supplies were rare, so it was in areas with more-
pronounced seasonal rainfall and less-reliable supplies that new generations of rulers assumed ever-greater power over water. Here, villages
tended to cluster around large-scale storage systems that provided the
water to nourish the field systems, where they grew crops not only for
themselves but also for the nearby lords. The rulers controlled the reservoirs, as well as large tracts of rich land immediately around population
centers. Their wealth enabled them to employ labor to maintain both
water systems and the field systems that depended on them. Places like
Tikal, in the Petén, which lies among some of the most fertile agricultural land anywhere in the tropics, became elaborate centers, surrounded
by an extensive hinterland of villages, most of them dependent, at least
in part, on the ruler's water.

Tikal and other major classic Maya centers came into their own after
250 C.E., when powerful dynasties of lords came to prominence.[10] Tikal
expanded dramatically during the first century B.C.E., as large public
buildings rose on the sites of humbler structures. Around 200 C.E., Yax
Ch'aktel Xok, or "First Scaffold Shark," founded a long-lasting dynasty.
He was the first of thirty-one rulers who presided over what became a
major city-state, whose domains may sometimes have included as many
as three hundred thousand people, sixty-five thousand or more of them
living in the immediate hinterland.

With its six temple pyramids, Tikal is a colossus among Maya centers.
The great plaza at the heart of the city is flanked on two sides by pyramids, and on the third by an imposing acropolis. Broad causeways link
outlying ceremonial complexes to the central precincts and served as
processional ways during important rituals. Tikal lies on the summits
of hillocks and ridges, so that the quarries for pyramid and temple
building laboriously excavated at their bases could become reservoirs.

The artificial hills and plazas of even small centers were symbolic depictions of the sacred landscape of mountains, hills, trees, and lakes, material replicas of the Maya cosmos designed as the settings for the elaborate public rituals that sanctified Maya life—and water management. Tikal, Belize's Caracol, and other centers were giant water catchments, their pyramids "water mountains." They were vertical water-management systems that moved water from higher to lower elevations, but under carefully administered conditions, even more controlled than those of the Balinese. During the rainy season, freshwater would flow down the steep sides and cascade into strategically placed reservoirs. During dry spells and the rainless months, the rulers would release water through carefully monitored channels into nearby field systems. Thus, in the immediate hinterland of Tikal and other such centers, control of water and its management lay in the hands of the ruler and a small nobility.

The small hills where centers like La Milpa, in Belize, and Tikal stood were engineered carefully for water runoff. They were "sculpted," as the Mayanist Vernon Scarborough puts it, heavily paved with plastered surfaces that sloped toward elevated reservoirs.[11] At Tikal, the highest,

Figure 13.3 *The central precincts of Tikal, Guatemala. (Tomfot/ iStockphoto)*

carefully engineered precinct covered 153 acres (62 hectares), a relatively impervious area that could collect more than 31,782,200 cubic feet (900,000 cubic meters) of water in a year when 59 inches (150 centimeters) of rain fell. This imposing catchment and others nearby filled reservoirs, natural depressions, and nearby swamps with enough water to last several years. The six reservoirs in Tikal's central precincts nestled against the gentle summits of small hills and contained between 3,531,478 and 8,828,668 cubic feet (100,000 and 250,000 cubic meters) of freshwater. The reservoirs lay in carefully arranged tiers so that water could be released downslope in a controlled manner. A network of raised causeways connected different areas and may have served as dams and boundaries for reservoirs. A simpler version of the same landscape had operated at Cerros about five centuries earlier.

Despite this centralized water management, Tikal's population lived over a wide area. Surveying great cities like this, shrouded as they are by dense forest, is a difficult undertaking, but so much work has been done on Tikal's hinterland, using both ground surveys and aerial photography, that we can estimate that 65,000 to 80,000 people lived within an area of about 46 square miles (120 square kilometers). A large ditch and natural swamps defined the margins of the hinterland, where as many as 650 people per 0.4 square miles (1 square kilometer) lived. Nearly everyone dwelled in outlying hamlets and villages, while relatively few people crowded into the immediate center of Tikal. The semitropical environment, with its seasonal rainfall and lack of permanent water, made true urban dwelling impossible.

The great lords who ruled Tikal and other major centers may have managed carefully engineered catchments and reservoirs, and the water that flowed from them, but they presided over a dispersed society, something very different from highly urbanized southern Mesopotamia. What developed was a much looser association than it might, at first glance, appear to be. Maya lords built imposing symbolic landscapes crafted from stone and stucco, but their domains were relatively small and constrained by poor communications and the environmental realities of the lowlands. Thus, Maya civilization of the first millennium c.e. was a patchwork of competing city-states, each with its own lord and subordinate rulers. Thanks to the decipherment of their glyphs, we know that the Maya world was a maze of diplomatic maneuvering, occasional warfare, and political marriages, where royal dynasties rose and fell with bewildering rapidity. As the Mayanist Lisa Lucero argues, these same lords acquired and kept the loyalty of their dispersed subjects by co-opting the powers of ancestor worship and ancient household ritual.[12]

THE SITUATION WAS somewhat different in those few places where water was plentiful, but even then the lords paid close attention to its management. Copán, in present-day Honduras, lies among the semitropical mountains in the southeastern part of the Maya homeland. Here, water is abundant, with plenty of springs and water holes, quite apart from the perennial Copán River and mountain runoff.[13] At least thirty natural water sources lie within the 9.6-square-mile (25-square-kilometer) Copán Valley. The muddy river water is undrinkable for much of the rainy season, but both rainwater and runoff were drinkable sources of still water when captured in reservoirs. Copán's major buildings suffered from poor drainage and flooding, and as early as 400 C.E., the builders laid out stucco and stone-lined drainage channels that ran among the major buildings. Much of the water flowed into nearby artificial ponds, which may also have yielded fish, mollusks, and edible and medicinal plants.

Copán's Acropolis was an elaborate water-catchment system, where internal drains caused sacred water to flow out of the mouths of sculpted earth and water deities on the façades of some buildings. The lagoons around the center were aquatic habitats for water lilies and edible plants, as well as tule rushes used for weaving mats and baskets. Copán's rulers wore the water lily headdress as an indication of their role in providing food and water.

Judging from the sculpture, a complex organization shared the tasks of water management in the Copán Valley. Each community in the valley may have been defined by its water holes, something reflected in some Maya hieroglyphic scripts, which refer to place-names and water locales, or "black holes." Thus, households and residential areas sharing the same water sources had a sense of shared social identity, as they do among some Maya communities today. At Zinacantán, in highland

Chiapas, Mexico, kin-based residential units share and inherit land. They form water hole groups that share and maintain a common water source. The members clean the water hole each year, share maintenance work, and perform the requisite offerings to the ancestors and water deities. An elected individual supervises the water hole and its rituals for the group. The same kind of organization may have flourished at Copán in ancient times. In a residential area known as Las Sepulturas, one building bears an elaborate façade in which the central figure is wearing a water lily headdress. He is flanked by two other figures wearing maize headdresses. Barbara Fash and Karla Davis-Salazar believe that these may depict representatives from individual kin groups seated beside a water hole manager. The façade highlights just how important it was for the Maya to manage and organize clean-water supplies in their reservoirs. At a higher level, there may have been some form of community council house, where representatives of the water holes, perhaps even of the rank of scribes, met to deal with larger-scale water-management issues. These became ever more important as the population increased and society became more complex. Above them sat the ruler, who, as an intermediary between the people and the ancestors and the forces of the supernatural world, was the ultimate political and ritual arbiter of water.

Palenque, in Chiapas, also thrived in a water-saturated environment, where the problem was not storing water, but getting rid of it.[14] The buildings lie at the foot of low hills covered with rainforest, just above the floodplain of the Usumacinta River. Palenque in its heyday was compact, constrained by the surrounding topography, and home to a dense population crowded into over one thousand structures. A labyrinthine building with a tower and interior plazas, known as the palace, dominates the site. The founders probably chose the location because of its abundant water supplies. Nine different watercourses, among them the sacred Otolum stream, flowed through the city. Instead of developing reservoirs to capture water, Palenque's architects constructed channels to carry it away from the urban landscape. In places, they built aqueducts, some of them underground, so that they could make efficient use of the limited area of flat terrain for plazas. Where Otolum's waters leave the city wall, an effigy of a caiman with jagged teeth stands

Figure 13.4 *The palace and observatory at Palenque. (Morgan Le Faye/iStockphoto)*

on the east side of the channel. Maya oral traditions tell us that a gigantic caiman lives at the center of the sky. He crouches expectantly. When a human prays for water, he opens his mouth and torrents of water escape. So the caiman at the end of the Palenque channel and aqueduct symbolically releases the water into the canyon for those who need it.

TODAY, PLACES LIKE Copán, Palenque, and Tikal are silent, monochrome ruin fields, where pyramids stand in silent majesty around empty plazas. We forget that they were once ablaze with bright colors, each of which had symbolic importance in a Maya world imbued with symbolism of all kinds. Each center was an axis mundi, a place where the World Tree linked the layers of the Maya cosmos. Each was also a symbolic replica of the Maya world, with its sacred mountains, trees, and open spaces. Tikal was a gleaming edifice, bright with brilliant colors and carved with hieroglyphic inscriptions. Then, suddenly, things changed. Tikal's population shrank dramatically after 830 C.E., to a mere 15 to 20 percent of its peak. Tikal was not alone: Dozens of other Maya city-states, large and small, imploded between 695 and 1050.

The causes of the so-called Maya collapse have attracted controversy for generations.[15] Some invoke social unrest and farmers' revolts. Others talk of disease, even disruptive volcanic eruptions. More recently, the Cariaco deep-sea core off Venezuela and borings into freshwater lakes have brought climate change into the collapse equation, for they have provided well-documented evidence of prolonged droughts that descended over the lowlands after 760 C.E. These droughts were connected to north-south movements of the Intertropical Convergence Zone, which brings rainfall to the lowlands when it moves northward over the Yucatán. If the zone lingers further southward, then drought settles over the Maya lowlands. There had been droughts around 150 B.C.E., at about the time when El Mirador was abandoned, but the Maya had recovered as new water strategies had come into use. Relatively wet times persisted between 550 and 750 C.E., when Maya populations grew rapidly and became much more vulnerable to dry spells. Both the Cariaco core and a boring from Lake Chichancanab, in the Yucatán itself, chronicle multiyear droughts that began as early as 760 C.E. and recurred at about fifty-year intervals—in 760, 820, 860, and 910. The arid cycles must have placed severe stress on major centers like Tikal, with their extensive reservoirs and almost total dependence on water storage. The pressures impinged on the lords, who always proclaimed themselves to be rulers descended from divine ancestors, with a special relationship to the forces of the supernatural. They were intermediaries between the farmer and the cosmos, the right order of things. Their authority depended on full reservoirs and punctual delivery of water to dry fields. When the royal reservoirs ran low, and the lords had to reduce allocations or were unable to provide any water at all, their credibility evaporated, probably with surprising rapidity.

The implosion of centers large and small was a complex process that was far from simultaneous. We cannot legitimately conjure up dramatic scenarios of thirsty farmers collapsing in their fields or in Tikal's plazas. Like the lowland's rainfall, the drought cycles had different effects even in contiguous areas, but we are still unable to study these highly localized events in any detail. Many Maya cities collapsed over a period of more than three hundred years through a concatenation of complex and still little-understood events. Maya rulers disappeared over a relatively short time between about 760 and 880 C.E.; remnant groups of elites and commoners stayed on at some centers. For example, Caracol, in Belize, was occupied through the 900s—but not by rulers; they were long gone, though the people were still there. There is no question, however, that water supplies played a critical part in the collapse equation, simply because many, though by no means all, Maya city-states flourished in areas with no perennial water supplies. The Cariaco core and its freshwater equivalents, as well as stalagmite data from the Macal Chasm cave, in western Belize, provide evidence for the four extremely dry periods after 760 C.E., but the problem is to find climatic data from individual sites.

The Maya reaction to prolonged drought involved not only major changes in settlement across the land but also numerous, still little-known water rituals. We know of one cult from remarkable discoveries in caves in Belize. The Maya rain god was Chac, who is shown on bark codices as residing in a cave and creating rainfall by pouring water from an upturned jar. Between 680 and 960 C.E., the Maya of western Belize created a water cult centered on this powerful deity. At Chechem Ha cave, Holley Moyes, Jaime Awe, and other colleagues mapped 984 feet (300 meters) of tunnels and the central chamber.[16] By plotting the distribution of charcoal specks from the Maya visitors' flaming torches and dating them, the team showed that the local Maya visited the central chamber to perform rituals around a central stalagmite and pool. They occasionally left potsherds behind. After 680, the visitors deposited more clay vessels throughout the cave, fifty-one of them complete pots, sometimes inverted, some in parts of the cavern never visited before. The Maya left more clay vessels behind them between 680 and 960 than at any other time. Many of the pots have wide mouths of a type most likely used for collecting water, just like those used by Chac. The excavators believe that the vessels were gifts to the rain god. The higher density of pots coincides with a lengthy dry period in the region, which is known from a stalagmite record in a cave only 9.3 miles (15 kilometers) away. When Moyes and Awe surveyed fifty-three other caves in central and southern Belize, they found large, intact jars lying on ledges and floors in many of them. They believe that they have recovered evidence of a widespread and hitherto-unknown drought cult, which coincides with the climatic data for arid conditions from paleoclimatic proxies.

There may have been another villain in some areas, too: deforestation. The Maya may have added to their suffering through their aggressive forest clearance. At Copán, in Honduras, pollen analyses have shown that the inhabitants had cleared at least 8.8 square miles (23 square kilometers) of pine forest by 800 C.E. Tom Sever of the University of Alabama in Huntsville has developed computer simulations of lowland deforestation using climatic data. Sever and colleague Robert Oglesby of the University of Nebraska–Lincoln ran two simulations—one of total deforestation, the other of a totally forested environment. A treeless environment would have led to a two- to five-degree rise in temperature and a 20 to 30 percent decrease in rainfall—much the same effect as deforestation in South America's Amazon Basin is having today. Settlement studies show that the hillsides were dotted with farmsteads—which means that land was cleared and soils eroded downhill. Of course, not all Maya land was deforested, so we need to look more closely at how farming and food production affected drought and temperature.[17]

When you examine the distribution of Maya centers across the lowlands, you add another dimension to the water equation. The largest and most prestigious centers lay in areas where farmers relied almost entirely on reservoirs. Where water was more abundant, centers tended to be smaller and the lords less powerful. No one needed lords, their social control, and their rituals, nor all the obligations of labor and food surpluses that went with them, when everyone had their own reservoirs. Nor did elaborate public ceremonies and communal water rituals make sense in an environment that revolved around the village and the household, where cherished rituals had persisted for generations. The Chac ritual in Belize may be connected more with villages than with larger centers and their lordly rulers, for a gift to the god in the form of a water pot is a personal donation, perhaps by a family or a household, which proclaims a much more personal relationship with the water deity than flamboyant rituals redolent of incense and flaring torches. Not, of course, that smaller centers and outlying communities were out of touch with the larger Maya world. Far from it, for exotic objects traded from afar have been found even in small villages, testifying to at least irregular contacts, both formal and informal, throughout the Maya world via a myriad of forest trails.

No question, the great Maya droughts had a profound effect on lowland society, especially on the lords who presided over paved precincts and sacred water mountains. Maya society was always dispersed, much of it dependent on local reservoirs and irregular water supplies. What the lords and their great pyramids brought to the equation in less well-watered areas was a predictability of water supplies that offered a measure of longer-term stability to large numbers of Maya households. The lords developed a powerful social contract between themselves and those they ruled, based on ties of ancestors and kin and reinforced by water rituals that they adopted for their own purposes. However, when the rains failed and drought settled over their brightly painted temples, the lords found that they had feet of clay. And in a world where loyalty was a fragile thing, based on such tangibles as gifts, protection, and predictable water, the villagers in their small communities exercised the ultimate power: They moved away, leaving their former masters to swing in the environmental wind.

The superstructure of Maya civilization survived for one thousand years, just as long as there was ample rainfall to fill reservoirs and maintain the myth of divine kingship, a special relationship to the supernatural that provided the water, which nourished the fields. After the droughts of the eighth and ninth centuries descended on the lowlands, much of Maya society returned to its much simpler village roots. The realities of irregular rainfall and only limited water storage kicked in. For the Maya, water was, ultimately, a local resource, because they lived in a tropical lowland environment that was effectively a green desert for at least four months a year. Land and water went together in a partnership that, in the end, defined the parameters of Maya society—and these did not involve the continued survival of powerful lords.

CHAPTER 14

Triumphs of Gravity

LAKE TITICACA, BOLIVIA: circa 800 C.E. Bone-chilling cold has descended on the altiplano (high plains) by Lake Titicaca during the winter night. White frost mantles the arid hillsides, where the local farmers plant their potatoes in thin soil. Many families watched all night as the icy air withered their crops. As daylight seeps across the hills, they wander through their ruined fields, staring down at a thin, white blanket of mist covering a plot of experimental fields on the plain close by. They have watched for weeks as archaeologists have dug up pampa sod, piled up layers of gravel, clay, and soil to form raised fields, and then dug shallow irrigation canals alongside them. Potatoes went in the plots; soon, green shoots grew higher than those on the hillsides. Then the frosts came. A white cloud of warm air hovers above the raised plots, hiding them from view. When the warming sun disperses the white blanket, the green potato plants are still intact, their leaves barely touched by frost . . .

Anthropologist Clark Erickson reconstructed a long-abandoned field system, one of the thousands of acres of such raised fields that once supported the powerful Tiwanaku state, which ruled a large tract of territory on the shores of Lake Titicaca, in the altiplano of what is now Bolivia, between about 400 and 1100 C.E.[1] The Tiwanakans used springwater and seasonal streams to water crops that fed thousands of people for centuries. With brilliant expertise, they even covered the foundations of some raised fields in marshy areas with a thick layer of lake clay to prevent the slightly saline water from Lake Titicaca from attacking the roots of the growing plants.

In the harsh landscape of the Bolivian altiplano, it was the same as in Sumer and in China—much water management lay in the hands not of autocratic lords but of the villagers. The work of building the raised fields was enormous, but the long-term rewards more than offset the initial labor. To what extent the polity of Tiwanaku was involved is a matter of discussion, but there must have been some involvement in the construction of large-scale waterworks in what was ultimately a managed landscape. Tiwanaku's highly successful water strategies stumbled in the face of prolonged droughts about one thousand years ago, and many fields were abandoned. Some local farmers have now started using raised fields anew.

The ancient Andeans thrived in desert and highland settings, even in places where there was almost no rainfall at all. Theirs was a world of contrasts, none greater than in the Amazonian lowlands, where Clark Erickson is now working in the Llanos de Mojos, in Bolivia, a place where the rainforest tapers into savanna.[2] Here, torrential rains flood tributaries of the Amazon, whose water overflows to the edges of circular patches, which Erickson calls forest islands, places where there was once ancient settlement. When the water recedes, it takes away the nutrients in the sandy-brown soil. Over thirty thousand forest islands are known, many hidden under the rainforest canopy. People have lived on the Llanos de Mojos since at least 1000 B.C.E. Researchers believe that early farmers managed the water and the forest. They would surround a suitable place with a large circular ditch. Then they would plant cacao trees, cutting down useless vegetation and cherishing the fertile brown soil. Chocolate was an important crop, its cultivation part of a huge ancient landscape of farming settlements, elaborate fish weirs, and raised fields linked by more than 6,000 miles (9,656 kilometers) of causeways. Through careful forest and water management, the ancient Amazonians contributed to the biological richness of their forested environment, and to its sustainability. When the first Jesuit missionaries arrived here in the late eighteenth century, they found what they called "civilized peoples," living in large settlements, who were famous for their chocolate beverages. Contrast this with the ruthless forest clearance of the slash-and-burn methods introduced by Spanish colonists, which have decimated the environment.

Erickson and others are pioneers in multidisciplinary historical ecol-

Figure 14.1 *Map showing locations in chapter 14.*

ogy, which studies the relationships between people and their changing landscapes. The Titicaca and Llanos de Mojos researches are examples of a new approach to ancient water management, which has important implications for a world grappling with water security. Human ecologists are realizing that the Andeans of centuries ago, whether on the eastern or western side of the range, were masters of hydrology, water conservation, and gravity-driven irrigation.

CHAVÍN DE HUÁNTAR, Peru: 900 B.C.E. Mountain rain mercilessly pelts the heads of the worshippers standing in the sunken courtyard. Mist and gray clouds press low on the ancient temple as the drenched crowd waits silently. The storm intensifies; water cascades into the honeycomb of narrow galleries inside the shrine. The sound of rushing water resonates through the hidden defiles, amplified by hidden, stone-walled chambers. A soft roaring echoes across the sunken court of cut and polished stone, reaching a sonorous climax, as the rain becomes a dribble, then a gentle mist. A seashell trumpet blares loudly. A masked, dancing shaman appears dramatically from the depths of the echoing shrine, chanting loudly. Deep in hallucinogenic trance, he delivers the pronouncement of the deity. The shaman vanishes as suddenly as he appeared. Drums beat; smoke rises high in the sodden air; the sound of the reverberating waters softens to a dull murmur and then stills. As the crowd disperses, the temple lies silent in the gloom . . .

Archaeologist Richard Burger believes that Chavín de Huántar was an artificial mountain where water rituals took place.[3] Ancient Andean beliefs held that the earth floated on a vast ocean. From there, water circulated through mountains to the Milky Way in the heavens. The gossamer-like Milky Way, or Mayu, the "Celestial River," was part of the critical barometer of heavenly bodies that measured the rhythm of Andean life. Andeans believed that Mayu carried water into the sky from the cosmic ocean that surrounded the earth. The Milky Way passed underground, took on terrestrial water that had flowed into the ocean, and rose again in the east. As the Milky Way moved slowly through the heavens, it deposited moisture through the celestial sphere, which fell as rain on earth and eventually flowed down rivers into the cosmic ocean, in an endless process of recycling. Thus, celestial and terrestrial rivers acted in concert to recycle the water that was the source of fertility, the subject of much traditional ritual in Andean communities to this day.

TWO GREAT POLES of Andean civilization developed on the western side of the great mountain range. One lay in the highlands, centered on Lake Titicaca.[4] The other lay to the northwest on the arid Peruvian North Coast. This coast is among the driest places on earth, virtually the only water coming from mountain runoff flowing down rivers dissecting the coastal plains. Here, as in other parts of the world, the earliest irrigation was probably on a small scale, the work of individual families and neighboring villages, but these tentative experiments soon developed into much more extensive systems. The coastal population rose rapidly. By 2000 B.C.E., some settlements near the Pacific housed between one thousand and three thousand people engaged in agriculture and anchovy fishing. These were close-knit communities where ties of kin and strong beliefs in the ancestors fostered a sense of common identity, reflected in flamboyant textile artistry—depictions of colorful anthropomorphic figures, crabs and snakes, and other creatures, preserved for us to admire in the coast's arid environment. Zigzagging serpents often symbolized water in Andean society, reflected in occasional snakelike irrigation canals that twisted and turned in seemingly impractical ways. The ceremonial maneuvering of water flow may have been thought of as a way of influencing natural rainfall and runoff.

None of the states that arose in the coastal river valleys ever achieved great size. The Moche, who dominated the coast between about 200 B.C.E. and 650 C.E., presided over a strip of coastline some 250 miles (400 kilometers) long, from the Lambayeque Valley in the north to the Nepeña Valley in the south. Most of the Moche lords' subjects lived in river valleys that fingered inland no more than about 50 miles (80 kilometers).[5] The lords probably acquired their domains in long-forgotten campaigns of conquest, but their enduring success came from control of valley water systems. The conquerors inherited a patchwork of village irrigation systems, some of considerable size, and linked them into networks of fields and irrigation canals. Moche villagers relied on highly flexible farming methods. They cultivated fertile plots on a small scale, laying them out along coastal hills where they could maximize runoff from springs and the occasional rainstorm. Such farming systems worked well when population densities were relatively low.

The state's agricultural base required relatively small-scale labor investment and no elaborate irrigation works. As in Mesopotamia, Egypt, and India, the focus of agriculture was the village and the dispersed local community. Much depended on mountain runoff during and after the rainy season, carefully shared between neighboring villages, that turned irrigable land into mosaics of green fields. The widely dispersed irrigation works gave a measure of protection against the sustained droughts and rare, epochal El Niño rains that occasionally caused landslides and could sweep away entire field systems developed over generations in a few hours. Moche farmers made use of every conceivable water source at their disposal, even growing crops in sunken gardens immediately above the water table. But the primary source of water was mountain runoff, a seemingly annual gift in the hands of the supernatural world. So Moche's rulers positioned themselves as intermediaries between the living and the powerful forces of the spirit realm, the providers of water. The lords' authority came from their perceived supernatural powers, reinforced by elaborate public rituals. They taxed their subjects in produce of all kinds and in compulsory labor, later known as *mi'ta*, which deployed large numbers of people to build vast monumental platforms and temples.

The Moche state eventually collapsed in about 650 C.E., to be followed by other polities. The greatest of them was Chimor, which came into prominence in the Moche Valley after 800.[6] Chimor's rulers, who presided over a much larger population than the Moche, lived in secluded compounds in their capital, Chan Chan. They invested heavily in closely organized, highly diversified agriculture and used *mi'ta* labor to deploy thousands of villagers on canal-digging projects and irrigation works. Chimor became a centralized domain based on agriculture and anchovy fishing administered by local nobles. Military force and a carefully administered tribute system, as well as efficient communications, ensured that everything flowed to Chan Chan, where thousands of nonfarmers dwelled, many of them skilled artisans in gold, silver, textiles, and clay.

CHAN CHAN: 1250 C.E., spring. The fog clings to the nearby coastline, the broad river valley a universe of nuanced gray. Men and women, huddled in simple capes, bend to their labors in the fields in the calm of morning. The ceaseless Pacific surf murmurs in the background. The workers advance slowly across the field as they dig into the damp soil

Figure 14.2 *Aerial view of elite compounds at Chan Chan, Peru.*
(*Charles Lenars/Corbis*)

before planting a new maize crop below the level of the river channel. Occasionally, they stand up and stretch, greeting neighbors working in nearby plots, whose heads and shoulders emerge suddenly from the fields dug below river level generations ago and harvested year after year. Everywhere they look, the landscape is arid and seemingly lifeless, except in the deep-set grids of fields that are invisible when they labor below ground level . . .

Most domestic water in Chan Chan came from step-down wells, especially near the Pacific, where the water table was close to the surface. Sunken gardens extended about three miles (five kilometers) inland, also taking advantage of the high water table. An enormous network of canals watered the flatlands north and west of the city. At the same time, the Chimu built redundant irrigation canals throughout their domain that supplied water to different parts of river valleys, some up to twenty-five miles (forty kilometers) long. This vast canal system was never in use all at one time, for there was insufficient water to fill all of it simultaneously. The communities that depended on the canals lived according to a carefully arranged timetable in which water was delivered to everyone at different times. Today, local farmers water their fields about every ten days; the Chimu probably did the same. As in Mesopotamia, much of the decision making about water distribution must have resided in local hands. In the Chimor case, the system was so large and varied that those relying on it could bring different parts of it into use if springs ran dry or El Niño flooding washed away some canals. In some places, elaborate overflow weirs formed part of the irrigation system, especially where aqueducts bridged large canyons. Stone-lined conduits allowed water to flow through the base of the structures without damaging them. Such measures provided some protection against catastrophic inundations.

While the Moche reacted to floods and drought by moving their communities from one location to another, the Chimu population was too dense to move. Chimor was a domain of large agricultural landscapes created with enormous labor, complete with storage reservoirs and terraced hills to control water flow down steep slopes. The authorities distributed water over long distances so people could grow two or three crops a year, where only one had been possible before—and that, at the time of the annual flood. Chimu rulers were ruthless in their administration. It could not have been otherwise, given their enormous investment in managing finite water supplies. They forced their subjects into large settlements and severely restricted individual mobility. Such draconian measures allowed them to respond to environmental uncertainty and major El Niño events on a regional scale, by bringing unscathed irrigation canals online and diverting crops from one area to another, while deploying large numbers of *mi'ta* laborers to repair aqueducts and canals within a short period of time. The Chimu survived major environmental disasters because they lived with drought every day of their lives and had the hard-won experience of their ancestors to draw upon.

The Chimu were a powerful force in the shifting political landscape of the Andes, but, inevitably, ambitious rivals cast covetous eyes on the wealthy state. In about 1470, Inca conquerors from the highlands gained control of the strategic water sources that nourished Chimor and incorporated Chimu domains into their growing empire.

THE CHALLENGES OF growing crops in North Coast valleys pale into insignificance compared with those of the people of the Nasca Valley, in southern coastal Peru, whose homeland lies in the bleak rain shadow of the western foothills of the Andes, on the northern fringes of the Atacama Desert, where no rainfall arrives whatsoever. Here, as along the coast to the north, rain falls in the Andes at altitudes over 6,550 feet (2,000 meters) above sea level. The rivers of the south produce a mere trickle in comparison with those of the north, and that for only a few months a year.[7]

The rivers that traverse Nasca flow down the western slopes of the mountains. Their courses are both shorter and straighter than those of the North Coast's rivers, and they pass through extensive deposits of volcanic ash that foster widespread seepage. By the time the rivers reach the coastal plain, there's insufficient water flow or velocity for them to run along the riverbed. Now the water only flows below the surface, in some places at depths of about thirty-three feet (ten meters). Almost invariably, the only available water supplies are underground.

The point at which the rivers disappear below the surface varies from year to year. When rainfall in the highlands is low and during droughts, the water vanishes at higher elevations, sometimes at altitudes as high as 4,900 feet (1,500 meters). In higher-rainfall years, the lowest the water comes before disappearing is about 2,624 feet (800 meters) above sea level. The rivers flow aboveground along the coastal plain only in years of exceptionally heavy rainfall, during February and March—perhaps in about two years out of seven.

The rivers then reemerge, usually at about 1,300 feet (400 meters) above sea level. Most human settlement and arable land lies well inland—in the case of the modern town of Nasca, about 37 miles (60 kilometers) from the Pacific. Nasca's rivers don't produce irrigation water every year, and when they do, the flow can only feed canals that extend between 0.6 and 1.2 miles (1 and 2 kilometers) from the source. To live and farm the most fertile soils of the middle valleys, the people have to dig for their water. The length of the Nasca Valley that is waterless is as much as 12 miles (20 kilometers). And yet, it was along this arid zone that human settlement concentrated, because of a remarkable local invention, the *puquio*, effectively a filtration gallery dug to capture subsurface water as it flows and seeps downstream.

We can imagine the puzzlement of the first Spanish conquistadors to arrive in the Nasca Valley. They would have gazed over green, well-watered fields in the midst of arid desolation, narrowing their eyes against the constant wind and flying dust. Turning their backs to the breeze, they would have looked for the rivers or wells that provided water. Instead, they looked down into shallow trenches heading upstream, ever deeper, V-shaped, here and there a villager cleaning out silt as water seeped around his feet. They learned that these trenches were called *puquios*. "Are these *qanats*?" they would have asked one another, familiar as they were with them from Spain.

Puquios share much in common with *qanats*, in that they are what archaeologists Katharina Schreiber and Josué Lancho Rojas call "horizontal wells." A *puquio* is not a tunnel like a *qanat*, but an open trench that taps into the water-bearing layer of a stream underground. Most are relatively shallow, the deepest reaching a depth of about thirty-three feet (ten meters).

As with *qanats*, much of what we know about *puquios* comes from modern practice, for they are still in operation. There's no reason to believe that their character has changed much over the centuries. According to Schreiber and Rojas, thirty-six *puquios* operate in three valleys, all oriented from west to east, to capture water from the natural drainage. They lie in the central part of the valleys, close to where the rivers vanish below the surface. The diggers create a V-shaped water channel, the base being about 3 feet (1 meter) wide. Stone cobbles line the base and the walls for a height of about 3 feet (1 meter) or so, the closely packed stones allowing water to percolate into the channel in the upstream segments. An earthen berm on the river side of the channel protects the *puquio* from floods. The trenches are shallowest upstream where

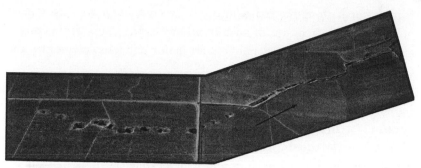

Figure 14.3 *A* puquio *from the air. Arrow shows the direction of flow.* *(Courtesy: Katharina Schreiber)*

the water goes below the surface and downstream where it reemerges. The depth of a *puquio* depends on the location of the water-bearing layer, and also on the gradient of the land surface. Most of them are between 1,640 and 4,921 feet (500 and 1,500 meters) long. The trench extends back upstream until it encounters subsurface water. At the downstream end, it flows directly into irrigation canals, or, more commonly, into a small reservoir known as a *kocha*.

When were *puquios* first invented? The first farming communities appeared here between 400 and 200 B.C.E., all of them established at higher elevations where perennial water could be found. The available water supplies were so sparse that there was, apparently, an increase in violence. We suspect this because settlements now tended to develop in easily defended positions. By 100 C.E., the Nasca culture, with its distinctive ceramics adorned with animals, plants, and mythic beings, was well established throughout the region.

Now a major ceremonial center developed in the intensely hot, sandstorm-ridden lower Nasca Valley. Cahuachi, with its dozens of pyramids and platform mounds, was the locus of an important burial cult and a major manufacturing center for spectacular polychrome pottery. Cemeteries extend for six miles (ten kilometers) along the lower valley, with Cahuachi at the heart of what was a giant necropolis. This was a special place, where only a few people lived permanently, perhaps the political and religious elite of Nasca society. Most likely, Cahuachi was a sacred location, a place of pilgrimage akin to others that flourished further north along the coast, in use between about 1 and 300 C.E., if not longer. Cahuachi was of such importance that people from considerable distances away gathered here for major religious ceremonies and to bury their dead.

Cahuachi is where the Nasca River emerges on the surface after flowing underground for more than 12.4 miles (20 kilometers). The comparatively well-watered location may have made this oasis a sacred place ideal for major rituals. There do not seem to have been very marked social differences between elites and regular villagers. Judging from archaeologist Kevin Vaughn's excavations at the village of Marcaya, some distance upstream, most people lived in small, self-sufficient farming communities where they relied on local water supplies.[8] Almost certainly, there was insufficient water for people to live in much larger settlements. Nevertheless, some form of centralized authority had come into being, with the clout to organize large labor forces to build and expand the platform mounds and pyramids at Cahuachi.

During the centuries when Cahuachi flourished, local people also surveyed and laid out many of the celebrated Nasca "lines," actually geoglyphs (drawings on the ground) that include both animals and plants. They lived on the edges of a desert that formed an enormous natural sketching pad. By sweeping away the loose surface stones, the Nasca created a web of white lines, some mere narrow tracks, others as wide as an airport runway. Some run straight for more than five miles (eight kilometers) across valleys and low hills. At ground level, they are difficult to discern, but viewed from high above the desert, the seemingly jumbled lines coalesce into vast geoglyphs of birds, monkeys, spiders, and plants. Aerial photographs reveal more than eight hundred miles (thirteen hundred kilometers) of straight lines, quite apart from figures. Several of the lines point at the mounds, cemeteries, and shrines of Cahuachi. The precise significance of the geoglyphs still eludes us, but they may be associated with local mountains, which figured prominently in rainmaking rituals, for they were the source of water for farming.

By the fifth and sixth centuries C.E., the Nasca people had begun a serious quest for more-reliable water supplies, perhaps in the face of prolonged highland droughts. Some communities moved to higher ground.

Many others did the opposite and relocated downstream in the dry middle valleys. The only way they could have done this was by tapping groundwater with *puquios*. The earliest archaeological sites associated with *puquios* date to about 450 C.E., just when the search for water was intensifying.

To develop *puquios* required no dramatic technological innovation, nor were they complex. Every farmer in Nasca knew well that rivers vanished underground and emerged or seeped out elsewhere downstream. They were also well aware that one could reach water by digging what was effectively a simple well. So to dig channels back toward the place where the water vanished was a logical step, especially for people with the surveying skills of the Nasca and given the seeming availability of considerable amounts of organized labor.

The droughts intensified between 550 and 750 C.E., events well documented in ice cores taken from Andean glaciers. The Nasca people continued to develop *puquios* as a result. Arable land abounded in the middle valleys, much more so than higher upstream, so there were strong incentives for continuing *puquio* use even after rainfall improved; they were, after all, the most reliable water supply around. *Puquios* now supported fewer settlements, but they were larger and more compact, with people living in much closer proximity.

In earlier times, irrigation was a simple matter of diverting water that emerged downstream into short canals that watered small plots of land. The digging of *puquios* changed the economic and political equation significantly. A small number of people organized their construction and controlled the water that flowed from them. They made decisions about water distribution beyond what they used for their own plots. Ties of kin and other complex social factors must have come into play, as water became a significant piece on the political chessboard, which was complicated in the first place. Inevitably, there were serious disputes over water that involved major changes in the economic and political scene. In places like Mesopotamia, such conflicts had led eventually to much larger political units—to cities and states. But in Nasca, as happened among the Hohokam of the southwestern United States and in other smaller-scale societies, the competition and conflict led to political fragmentation, to a world of smaller polities, each with its own rulers, its own water supplies, and the instinct to compete with its neighbors. Cahuachi still received the dead, but it was no longer an economic or political center, just a place of veneration.

The later history of Nasca was convoluted, as its farming communities became part of the highland Wari Empire after 750 C.E. The Wari built one center at a lower elevation, watered by a *puquio*, and several others in the upper reaches of the drainage. Their domain collapsed around 1000 C.E., and Nasca resumed its independence, with an almost total dependence on *puquios*. So matters remained until the Inca incorporated the Nasca into their enormous empire. Interestingly, the Inca never established an administrative center in Nasca. Only one lordly estate is known, which lies in the Nasca Valley close to land watered by two *puquios*. Both are impressive facilities that are still in use. One of them, La Gobernadora *puquio*, originates 26 feet (8 meters) below the surface and now passes through a 1,220-foot- (372-meter-) long tunnel before opening into a channel with an aggregate length of 997 feet (204 meters).

The arrival of the Spanish had no initial effects on *puquio* use. In time, Spanish colonists realized the advantages of such water devices and expanded those close to their towns, sometimes with lengthy tunnels and even branches, to increase water flow. By the nineteenth and twentieth centuries, wealthy landlords controlled most use of *puquio* water, which they employed for their large haciendas. They built their ranch houses close to the *kochas*, which provided them with a reliable water source. They also converted many of the old open trenches into covered galleries in order to increase their agricultural lands.

Today, *puquios* are a dying technology in the face of earthmoving machinery and diesel pumps. A modern version, known as a *pozo-kocha*, dispenses with the long channel, a pump now extracting the water at the upper end. More groundwater comes to the surface, but the water supply for traditional *puquios* diminishes as the filtration layer goes further underground. *Pozo-kochas* consume much less farmland and are easier to maintain—if you have the money and the fuel to run them. *Puquio* owners now complain that the water table is dropping, as promiscuous pump use is overdrafting the subterranean water source. What the future holds is uncertain, for here, as elsewhere, pumping groundwater probably has no sustainable future, as opposed to the centuries-old, self-sustaining *puquio*, which turned extensive tracts of Andean desert into verdant farmland.

THE INCA WERE masters of water engineering and had the wisdom to leave Nasca's *puquios* alone. Their realm, Tawantinsuyu, the "Land of the Four Quarters," was a vast, closely organized empire, its hub Cuzco in the highlands, where the genius of the Inca's water engineers, and probably those of conquered Chimor, came into full play.

Cuzco was a city of single-story houses with steeply pitched thatched roofs. A stone-lined channel of fast-flowing water that ran down the middle of each paved street provided better sanitation than anything known in Europe at the time. The water was so clean in the two small rivers running through the city that the Inca bathed there. One flowed through the central plaza, on the eastern side of which lay the ruler's imposing palaces and ceremonial buildings. Most of Inca Cuzco is gone, but we are fortunate to have a portrait of Inca hydrology from two locations: Tipon and Machu Picchu.

Tipon, near Cuzco: 1400 C.E., spring. A small group of water engineers cluster around a flat boulder in the morning sunshine. Two of them bear quipus, the ubiquitous knotted strings used by Inca scribes as inventories. They are calculating potential maize yields from the ravine below them. Meanwhile, a finely dressed engineer watches his assistants as they puddle fine clay and deftly create a model of the valley and surrounding hillsides. Carefully, they mark the position of the hillside spring above the ravine and the course of the nearby Río Pukara. The engineer tells them to make another model, this time with the valley filled in from the hillside to the east. He marks off a series of terraces on the fill, plots in canals, develops a mental plan of the great estate within the already existing stone enclosure wall . . . This anonymous engineer and his peers created a hydrological masterpiece, a triumphant use of gravity.

Figure 14.4 *Tipon. (Courtesy: Kenneth and Ruth Wright)*

Tipon lies thirteen miles (twenty-one kilometers) downstream from Cuzco, a 500-acre (202-hectare) self-contained and walled estate that may have belonged to the ruler Viracocha Inca in 1400.[9] The estate is in a ravine, at its head the perennial Tipon spring. It results from the same kind of geology that provided water for ancient Greek cities: a solid volcanic layer overlying permeable limestone strata.

Viracocha's engineers began by filling in the valley. They piled boulders and rock fill, then smaller cobbles, gravel, and sand to form a huge natural reverse filter. Then they built thirteen terraces that extend 1,300 feet (396 meters) up the ravine, filling them with rich topsoil in which the estate grew maize, flowers, and herbs—prestige crops. The massive stone walls of the terraces resisted soil pressures and earthquakes, the irrigation water and rainfall percolating effortlessly through the natural reverse filter without adding hydraulic pressure to the walls. Near-vertical defiles, or water drops, built into the terraces controlled the water flow of the spring upslope.

The terraces, carefully integrated one with the other, act like a stairway down the filled-in ravine. The noble residences, a two-story grain store, and a military facility overlook them. A long irrigation canal

Figure 14.5 *Terraces at Tipon. (Courtesy: Kenneth and Ruth Wright and the American Society of Civil Engineers)*

from another valley surrounds the terraces on three sides. A large cere-monial plaza, an aqueduct, and a religious complex lie 600 feet (182 meters) to the northwest. The nobles and their retinues of estate man-agers and craftspeople dwelled in a small community of about one hun-dred people, immediately to the north. Every arable part of Tipon was terraced for cultivation, for either irrigation or dry agriculture. A mas-sive defensive stone wall surrounds what was a substantial community, rising to a height of 15 to 20 feet (4.6 to 6 meters).

Flood irrigation allowed the estate farmers to grow two maize crops a year, one harvested in January, the other in July. By using canals, the water managers could distribute water to the entire terrace system, as well as to the palace and ceremonial plaza area and the dwellings nearby. The Tipon spring produced about 300 gallons (1.4 cubic me-ters) of water a minute. The builders faced the spring with a stone wall and three spouts that enabled a user to draw water into a pot.

Downslope, the stone-lined discharge channel bifurcates and bifurcates again, which enabled the managers to guide water in different directions. The carefully engineered canals and their drop structures ensured an even flow.

The uppermost two central terraces lie above the Tipon spring. They received water from a side channel off the main canal, which brought water from the perennial Río Pukara. Three irrigation canals carried water from the river, about 0.84 miles (1.36 kilometers) north of the central terraces. The main channel followed the contours of the land and provided water to surrounding agricultural tracts, at times dropping sharply down steep slopes, with an average gradient of about 10 percent. Next to the ceremonial plaza lies a 200-foot- (61-meter-) long boulder aqueduct standing 15 feet (4.5 meters) above the ground, with a pedestrian underpass that also released floodwater.

Most irrigation systems are utilitarian at best, but to the Inca such works were also symbols of human control over water. Tipon was an enclave of privilege, where the ruler and his nobles could live in comfort to the sound of soothing water that flowed day and night and never ran short. At the same time, the engineers built for the long term, blending their canals and water drops effortlessly into the natural environment in an ultimate display of the power and use of natural gravity.

TIPON WAS BY no means unique; Inca water management was always efficient and devised for the long term, even with routine agricultural terracing. However, the showpieces were the royal estates and Cuzco itself. The most famous is Machu Picchu, the estate of the Inca ruler Pachacuti Inca Yupanqui, which lies high in the Andes at an altitude of 8,000 feet (2,430 meters) above sea level, fifty miles (eighty kilometers) northwest of Cuzco.[10] The well-preserved site is on a high ridge between two conspicuous peaks: Machu Picchu and Huayna Picchu. The fast-flowing Urubamba River, 1,640 feet (500 meters) below, surrounds it on three sides. Between 1450 and 1540, up to a thousand people lived at Machu Picchu when the ruler was in residence, about three hundred when he was not.

Figure 14.6 *Machu Picchu. (Jarnogz/iStockphoto)*

The location may have been chosen because of its position relative to nearby sacred peaks, but it was also feasible because of a perennial spring that emerged from the north slope of Machu Picchu. Seventy-nine inches (almost two hundred centimeters) of rainfall annually also provided water for agriculture. Pachacuti's engineers captured the springwater by setting a permeable stone wall into the steep hillside of Machu Picchu mountain. The springwater percolated through this humanly made filter into a collection trench about 2.6 feet (0.8 meters) wide. Water flow varied through the year, from about 6 gallons (23 liters) to 33 gallons (125 liters) a minute, depending on seasonal rainfall.

Water traveled 2,457 feet (749 meters) from the Machu Picchu spring down a small cut-stone canal built atop a carefully graded terrace above the terraced agricultural area. The canal itself was between 20 and 26 square inches (125 and 168 square centimeters) in cross section, with a meticulously varied gradient, so much so that a 1 percent gradient above

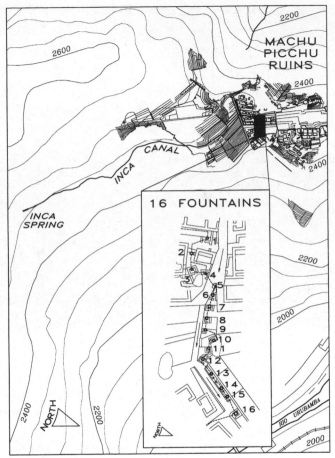

Figure 14.7 *Plan of the Inca canal and fountains at Machu Picchu.*
(Courtesy: Kenneth and Ruth Wright and the American Society of Civil
Engineers.)

the agricultural terraces slowed the flow and allowed surplus to spill out
onto the terraces. Machu Picchu's water engineers took care to maintain
the purity of the domestic water supply by directing agricultural and
urban storm-water discharges away from the spring canal.

The canal traveled to the doorway of Pachacuti's residence, where a
fountain channeled water into a rectangular basin cut into a rock slab.
The Temple of the Sun, built with great care, stood nearby, where a sacred
fountain with carved stone blocks and niches for offerings lay adjacent to

a sacred stone (*huaca*). The sound of plashing water may have played a role in religious ceremonies. The springwater now flowed downslope adjacent to a long stairway. Here, sixteen fountains, each with a spout causing water to fall into a stone basin, and placed in series over a vertical drop of about eighty-five feet (twenty-six meters), provided domestic water supplies for the residential areas. Stairways led to each fountain, except for the lowest, reserved for the Temple of the Condor. Surplus water flowed from there into Machu Picchu's main drain.

Just as in Roman baths, the public springs must have been centers of gossip, of social interaction. We can imagine groups of women and girls hefting *aryballos* (clay water jugs), filling them under the flowing spouts, then carrying them carefully along narrow walkways to their homes. No one is in any hurry, for the waiting and conversation is a highlight of the day . . .

Hydrological engineer Kenneth Wright estimates that the basic domestic water needs of the residents would have required that about three gallons (ten liters) a minute be delivered to the fountains. Backup supplies would have come from the Urubamba River, far below the community, an arduous climb down a trail that ended at the water. Additional springs and fountains lay along the trail, nourished by groundwater and accessed by tapping into the seep areas on the slope.

Machu Picchu's agricultural terraces, with their stone retaining walls, covered twelve acres (five hectares), hugging the contours of the terrain in a way that caused them to melt into the natural topography. Wright aptly describes them as being "like a quilted blanket laid out over the ridge."[11] Each was carefully constructed with loosely packed stones and chips from stonecutting overlain by more-granular soils, then thick, sandy loam topsoil. Carefully laid-out channels carried excess runoff away to nearby drains. The terraces themselves were never irrigated, for there was sufficient rainfall for dry agriculture. Crops of maize and assorted flowers and herbs came from the terraces annually, but not enough food for everyone. Machu Picchu relied on imported foodstuffs.

Machu Picchu's drainage system included a main channel that separated the agricultural and urban sectors, the latter covering about 21 acres (8.5 hectares) and comprising about 170 thatch-roofed buildings. The runoff from buildings and plazas either soaked into the ground—plazas lay atop layers of rocks and stone chips—or flowed into drain outlets placed in retaining and building walls. The Inca intercepted surface drainage, controlled small landslides, and delayed constructing terraces until they were certain the movement had ceased. They built everything for permanence and durability, for longevity, taming gravity and using it to sustain farming, a royal household, and up to one thousand people for nearly a century.

Inca hydro-engineering and water management was achieved by hand labor and with the simplest of stone and copper artifacts. Today's engineers have much to learn from them about protecting steep slopes from erosion, handling water in steep terrain, urban drainage, and building for the long term. As Wright remarks, "It seems the Inca would have had the equivalent of an urban drainage manual similar to the ASCE [American Society of Civil Engineers] Manual of Practice . . . except that the Inca had no written language."[12] They harnessed gravity triumphantly.

PART V

Gravity and Beyond

From gravity to deep pumping: In which we explore how ancient water management based on gravity reached brilliant heights in the diverse arid and semiarid environments of the Islamic world. The legacies of this achievement, and of Roman hydrology, pass into medieval practice, and technological innovations, including more powerful waterwheels and steam power, arrive with the Industrial Revolution. Chapter 17 explores the implications of our unrestrained mining of hitherto-inaccessible aquifers as we enter a new stage in the human relationship with water.

CHAPTER 15

The Waters of Islam

"WITH WATER WE MADE ALL LIVING THINGS," states the Holy Quran. The Arabic word for water, *ma'*, occurs sixty times in its pages and that for rivers on over fifty occasions. Water is a sign of Allah's mercy: "It is Allah who drives the winds that raise the clouds and spreads them along the sky as He pleases and causes them to break up so that you can see the rain issuing out from the middle of them." He sends water to barren lands to bring forth crops and "leads it through springs in the earth." The four rivers of paradise are incorruptible, part water, part "rivers of milk of which the taste never changes, rivers of wine, a joy to those who drink, rivers of honey, pure and clear."[1]

Such is the celestial paradise. Paradise, to those who lived in the hot, often dry, harsh environments of the Islamic homeland, was a green oasis, a place where a spring welled forth miraculously from the earth, the ideal of serenity and well-being. A classic Arabic adage lists three things that gladden the heart: "water, greenery, and a beautiful face." Countless Islamic poets have extolled the virtues of water as a merciful gift in the form of dew and rain, a river or a spring. Many of them wrote of a quest for the Water of Life, which bestowed immortality or restored youth. Water reflected the dual natures of Allah: violent storms and floods brought forth in anger and aimed at the "iniquitous and un-faithful," and gentle rain and ample water provided as acts of mercy. Water was never an entitlement and meant far more than drinking or irrigation. "When you rise to pray, wash . . . If you are polluted, cleanse yourselves." Ceremonial ablution (*wudu*, washing before prayer) and purification before prayer remain a central part of Islamic religious practice. The Prophet himself compared daily prayers to the cleansing action of water in one of his hadith.[2]

The ancient Islamic world was famous for its wonderful gardens, created with symphonies of effective but simple waterworks—barrages and canals for diverting water, wells and *aflaj* (*qanats*; singular *falaj*) to tap groundwater, well-placed water jets and reflecting pools built in the hearts of palaces. Near Córdoba, the great city of al-Andalus, in what is now Spain, a renovated Roman aqueduct carried water down from the mountains, flowing into the north side of the Madinat al-Zahra palace, then through the halls and gardens, before filling a large fish tank, smaller pools, and an ornamental basin in the baths next to the main reception hall. A large basin featured croaking tortoises and water pouring from the jaws of a lion fountain carved in amber and adorned with a pearl necklace. Elsewhere, water lilies grew in large ponds. Such gardens soothed the senses, provided serenity in a harsh, dry world. But they were far more than mere diversions. They celebrated the blessings and meanings of divinely given water, provided spiritual solace, were sources of profound aesthetic satisfaction. Cool shade, calming scents, the lushness of the plants, and, above all, the gentle symphony of running water—these were the magic elixir of Islamic gardens. Whether private or public, gardens were a glimpse of the paradisial, heavenly gardens of the next life.

Effective water management was a catalyst of the Islamic world and forms a significant part of Islamic law. The Quran warns against hoarding water; the Prophet taught that on the day of resurrection Allah would ignore those who possessed surplus water and withheld it from travelers. A hadith proclaims that all Muslims have a common share in pasture, water, and fire. There are well-defined water rights. The first is the law of thirst, the right of humans to quench their thirst; the second is the right of cattle and household animals to do the same; and the third is the right of irrigation. Water is also a gift for nourishing vegetation. At the same time, the Quran points out that God provides fixed amounts of water, which means that supplies must be managed and never wasted. The Prophet himself washed with a mere two tenths of a gallon (three quarters of a liter) of water and adjured against wasting water even when performing ablutions alongside a fast-moving river.

With its strong emphasis on conducting affairs through mutual consultation, a great deal of Islamic water management in medieval and later times unfolded at the local level. Under this rubric, water was a social good, a gift from God, its use governed by the water laws in the Quran. Islamic teachings placed a strong emphasis on stewardship, on conservation of water supplies in an equitable, sustainable way, acknowledging its central place in both secular and spiritual life.

DURING RAMADAN IN the year 610 C.E., forty-year-old Muhammad had the first of a series of electrifying revelations in a hillside cave on Mount Hira, near Mecca.[3] The message that emerged was simple: Charity, tithing for the *ummah*, daily prayer, and the *shahada* that proclaimed, "There is no god but God and Muhammad is his prophet." The Prophet's hegira to Medina (Yathrib) from Mecca in 622 culminated in

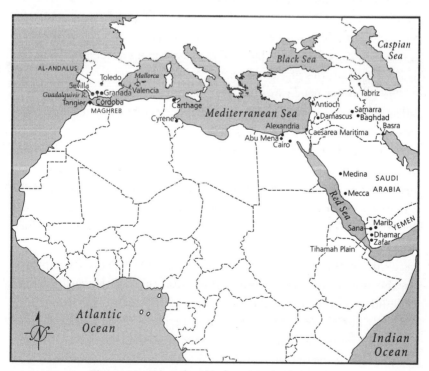

Figure 15.1 *Map showing locations in chapter 15.*

his triumphant return to his home city eight years later. Mecca's Ka'bah ceased to be a pagan shrine and became the spiritual epicenter of the new religion.

This momentous event transpired below the grandiose political radars of the Greco-Roman and Iranian monarchs, who were engaged in total war, as they had been for generations. But Mohammad's legacy would outstrip that of any conquering king. The Prophet died in 632 without naming a successor, but his religion continued to expand. By 656, the Quran was in its final form, with its 114 suras. The "guide for the righteous," whether read or recited, provided a compelling message that transcended tribal loyalties and forged all believers into a much greater tribe, the greatest of all: Islam. Jihad, a spiritual struggle in the path of Islam, became an irrepressible force. Less than a century after Mohammad's death, a huge, loosely knit Islamic zone of commerce, finance, and trade extended from Samarkand, in central Asia, toward the borders of China, and from Sind, in western India, to Tangier, on the Atlantic coast of North Africa. The warriors of jihad swept aside empires and kingdoms with astonishing speed, partly because of religious zeal and a desire for riches, but also because Islam's adversaries were exhausted by centuries of vicious wars.

The conquest began in 636. Syria, Palestine, Egypt, and Libya had fallen like dominos by the early 640s. By 637, the Sassanian Empire was mostly in Islamic hands. As the Islamic Empire grew, Damascus became an imperial capital under the Umayyad caliphate, with its main street running nearly 5,000 feet (1,524 meters) from east to west. The Aramaeans, Semitic nomads from Mesopotamia, had long ago established the water system for the city, constructing canals and tunnels to exploit the flow of the Barada River, which watered the Ghouta Oasis, in which Damascus lay. The city also drew water from the Fijeh spring, on the west side of the river. Both the Romans and the Umayyads improved on the system, the latter to such effect that the eighth-century caliph al-Walid proclaimed that the city acquired its superiority from "your climate, your water, your fruits, and your baths."[4] Damascus was, of course, far more, the center of a free-trade zone that extended the length and breadth of the known world. Meanwhile, conquests continued. The eastern Roman empire and the Berbers of the Maghreb, "the place of sunset," in northwestern North Africa, maintained their autonomy, though the latter for only a short while. By 690, Carthage was in Muslim hands despite serious reverses for Islamic warriors. Adept politicking as much as warfare won over the Berbers, many of whom became administrators and warriors in the service of the caliphate. Musa ibn Nusayr, a shrewd general and an even smarter politician, completed the subjection of the Maghreb and, in 711, appointed one of his subordinates, Tariq ibn Ziyad, to probe the lands visible across the Strait of Gibraltar, soon to be known as al-Andalus. By 714, most of Spain as far north as the Ebro River and west to the Atlantic in what is now central Portugal was no longer in the hands of the Visigoths. Al-Andalus was to become a magnificent center of commerce and learning, sustained by water-management expertise acquired from many lands.

THE ROMANS CALLED Yemen Arabia Felix (Happy Arabia) on account of its precious frankincense and myrrh, as well as its wide expanses of relatively well-watered highland topography.[5] Yemen, like the rest of the Arabian Peninsula, was, in many respects, a nursery for water-management expertise in arid and semiarid landscapes. The methods varied from place to place, relied on the simplest technologies, and changed imperceptibly over the millennia. Four, sometimes five, powerful kingdoms flourished simultaneously at intervals after 1200 B.C., all located on the edge of the Rub' al-Khali (Arabian Desert). All prospered in part because of their trading skills and in part because of their careful water-management and farming practices, but in general water was a local matter, in the sense that supplies were localized and finite, often managed by small tribal societies whose survival depended on efficient harvesting of irregular water for their crops and flocks. Agriculture was the backdrop, but international trade along caravan routes that passed around the central desert was an important catalyst for political power.

The spectacular landscapes of Yemen encompass some of the driest deserts on earth, as well as arid lowlands and relatively well-watered highlands more than 2,500 feet (762 meters) above sea level. Long before Islam burst upon the scene during the seventh century, farmers and

herders had made expert use of sparse water supplies to support themselves in what were often marginal environments. They had made use of
simple water-management practices, harnessing runoff as it cascaded
down wadis large and small, terracing hillsides, and feeding them with
water stored in numerous reservoirs. Such methods were in widespread
use in Yemen by the fourth or third millennium B.C.E., at a time when
population densities were low. We know that sophisticated flood-control
systems based on earlier harvesting techniques have operated on the
fringes of the desert since at least 2000 B.C.E. and remain in use today.

Between the tenth and second centuries B.C.E., the wealthy kingdom
of Saba (or Sheba) prospered off the spice and incense trade. The Sa-
baean capital, Marib, lay on the edge of the great desert and controlled
the frankincense route that ran from eastern Yemen and Oman through
the Arabian Peninsula to the Mediterranean. Saba, with its legendary
associations with the fabulously wealthy Queen of Sheba mentioned
in the Old Testament and the Quran, was also an agricultural powerhouse, thanks to its efficient irrigation systems. Some 19,700 to 23,700
acres (8,000 to 9,600 hectares) of irrigated fields provided food for the
247-acre (100-hectare) town. The irrigation water came from the 2,230-
foot- (680-meter-) long Marib dam, first constructed during the sixth
century B.C.E. by two rulers of Saba. The dam was the culmination of
the so-called *aqm* systems of flood irrigation that had been in use in Yemen since at least the third millennium B.C.E. As we have seen, such systems divert floodwaters with stone, earth, or brush structures that guide
the flow to fields located progressively downslope. Almost certainly, local water masters and landowners administered the *aqm* systems, the
former being paid in kind. Letters and inscriptions by water masters
written during the late second and early third centuries C.E. preserve
details of water transactions and offer thanks for uneventful terms of
office in what must at times have been a fractious job. However, dams
and irrigation systems the size of Marib's required a more collective approach to management that would certainly have come into play when
exceptional floods breached a dam or canal works. Given the region's
volatile, highly factionalized tribal groups, decisive leadership was required to navigate local feuds and organize communal labor when
needed. The resources assembled to repair a major dam were on an im-

pressive scale. According to a Yemeni inscription hewn into the stones at Marib and dated to 449 B.C.E., fourteen thousand camels were needed to strengthen the Marib dam in that year. A year later, it collapsed. We are told that the ruler of Marib, in an impressive display of charisma, rounded up about twenty thousand tribesmen to rebuild it.

Saba gradually declined after the fifth century B.C.E., as competing kingdoms came into prominence. The discovery of the monsoon winds of the Indian Ocean by Egyptian sailors took the lucrative spice trade offshore, and the caravan cities lost much influence. By the second century B.C.E., the Himyarite kingdom dominated a large area, conquering Marib and becoming a powerful land and sea power for four centuries.

Working in the expansive Wadi Dhanah region, British archaeologist Tony Wilkinson and a team of researchers have studied water management in a 3,860-square-mile (10,000-square-kilometer) catchment of the Yemen highlands.[6] They found three systems in use. In the highlands and other better-watered areas, farmers built extensive terrace staircases going back to at least 3,000 B.C.E. Such terrace systems, watered by rainfall and runoff, were the work of individual families, or groups of families, who also maintained small-scale floodwater farming operations. In many cases, the labor of terrace building was not enormous, for sediment washed down from higher elevations helps develop the fields. Elsewhere terrace building often required the cooperative efforts of large numbers of people. These highland terrace systems were, and still are, highly flexible and can be extended wherever there is suitable land. For all this flexibility, each community administers them scrupulously, using comprehensive local laws to ensure utter fairness and proper use. The resulting irrigation works have been self-sustaining over many centuries, as were many small-scale water-management systems throughout early Islamic domains.

Drier parts of the highlands, with annual rainfall in the 7.8- to 13.7-inch (20- to 35-centimeter) range, can also support terracing, but the fields require careful placement so that the correct moisture levels can be maintained by catching runoff from upslope. In general, the drier the landscape, the larger the catchment area needed to water ever-smaller areas of productive land. The driest areas rely on the runoff from entire wadis, where small dams or earthen embankments deflect floodwater,

Figure 15.2 *Terracing at Dira, Yemen. (Michel Gounet/Godong/Corbis)*

then distribute it to individual farmers. Such a flood-irrigation system (*sayl*) still operates on the Red Sea's hot and harsh Tihamah coastal plain, which runs from Saudi Arabia all the way down to southern Yemen, the temporary barriers being maintained by those who hold land irrigated by them. The heads of local communities organize and administer *sayl* irrigation in both Saudi Arabia and Yemen, as they have for at least two thousand years.

During the first millennium B.C.E., the farmers of the Yemeni highlands around Dhamar relied on terrace agriculture and rainfall, supplemented by irrigation water stored behind stone-built dams. Distribution channels carried the water to valleys and lowlands downstream. How long the dams were used for actual storage is unknown, but they were certainly employed for this purpose in the short term. Such barriers, often of impressive size, were vulnerable to unusually high floods. The Sedd al-Ajmar dam is about 296 feet (90 meters) long, up to 49 feet (15 meters) thick, and 46 feet (14 meters) high and once captured water, along with another dam, from two wadis. However, a major flood eventually breached it and rendered it useless in a few hours. During the time of the Himyarite kingdom (circa 110 B.C.E.–sixth century C.E.), modules of dams formed an important component in large-scale irrigation systems, even in areas with more abundant rainfall. Such modules could not supply the entire population, but provided water for numerous small towns or villages, despite dam failures; nearby terrace systems provided a safety net.

Around the Himyarite capital of Zafar, wealthy from the east African ivory trade, densely packed irrigated fields occupied the valleys, passing into terrace staircases up adjacent hillsides. The valley floors depended on dams and basins that watered interlocked modules of fields, each covering about 24.7 acres (10 hectares), enough to feed a large extended family. Some eighty dams are said to have provided irrigation water for the capital and its surrounding lands, despite the relatively well-watered landscape. This enabled the authorities and local farmers to produce multiple crops each year rather than making up for deficiencies in rainfall. As so often happens to dams, most of Zafar's are now silted up.

The scale of Yemeni irrigation systems was much larger on occasion, requiring stronger, more centralized leadership, but there was none of the highly organized control typical of Sassanian water management. There was a basic continuum from rain-fed terrace agriculture maintained by families in the highlands to small-scale runoff systems in the semiarid desert margins, again run by small social groups. There were also much larger systems managed by rulers of communities with more than ten thousand inhabitants. There was, and still is, an impressive flexibility in Yemeni water management that surely reflects what was the case over much of the ancient Islamic world.

These various traditions, practiced over a wider area than just Yemen, as well as long-established methods of tapping groundwater, such as wells and *aflaj*, traveled far beyond the confines of the Arabian deserts as Islamic jihads unified enormous areas of the known world.

THE GREAT CONQUESTS of the seventh and eighth centuries C.E. brought an enormous tract of the world under a single language, a single religion, and a common legal system. Islam's empire was at first a loose alliance, which depended on allegiance to the Prophet. A relatively strong and durable state emerged under the Umayyads and, later, the early Abbasids of the eighth century. The caliphate spanned a highly diverse semiarid and arid heartland, as well as encompassing monsoon lands in Yemen and northwestern India and near-temperate and continental environments in al-Andalus and central Asia. The Umayyads and the Abbasids inherited or forged far-flung trade links, along Indian Ocean water routes and over thousands of miles of desert caravan tracks and roads that connected growing cities, major markets, and ports. Within Islam's vast domains lay numerous, often rich cultural traditions that had developed over thousands of years, and the societies that had built them had much to learn from one another. The Islamic Empire had footholds in three continents—in Africa, Asia, and Europe—so much so that Muslim scholars acquired a unique knowledge of all parts of the known world. Their learning dwarfed that of the West for centuries.

The fourteenth-century scholar Ibn Khaldun wrote of the Arabs that "all their customary activities lead to travel and movement." Theirs was a world of perpetual motion of people, goods and commodities, and information and ideas. The conquest resulted in wholesale population movements and in the settlement of conquering soldiers in new lands. Within a few generations, an astounding mixture of different peoples inhabited the towns and cities of Islamic domains, many of them having come from places as far-flung as India and Yemen. Pilgrimages added to the constant restlessness, attracting large numbers of the devout to major shrines, especially to Mecca. Wrote geographer and poet Ibn Jubair of Mecca, "There is no merchandise or precious object on earth which is not to be found in this town at the time of the pilgrimage."[7]

New crops spread across the Islamic world in the wake of conquest and scholarship, which led to a wider knowledge of Indian medicinal plants in particular, as well as foods like rice and sugar. The trade in medicines and imported crops may have been confined at first to the wealthy; exotic foods like sugar and sour oranges starred in elaborate feasts. A taste for such foods spread rapidly westward, and others sought to emulate these displays of wealth. Inevitably, if gradually, a growing demand for such crops as cotton and rice led to local experimentation and, after much trial and error, the routine growth of formerly exotic foods. Cotton, originally an Indian crop, offers an excellent example.

The Persians adopted it and traded in textiles as far away as Egypt. By the tenth century, the Egyptians were growing cotton for their own use. Thirteenth-century cookbooks for wealthy families in Baghdad and al-Andalus boast numerous recipes for such exotics as eggplants, sour oranges, and lemons. Sugar was a commonplace flavoring.

All of these crops, and swelling populations, required much larger water supplies in landscapes where water was always scarce. Many of the new crops were native to environments like India's Indus Valley, where there was a season of heavy rains. In contrast, both the Near East and the Mediterranean world enjoyed much lower, unpredictable rainfall, much of which fell in the colder winter months. Most crops required irrigation during the dry summer, and often for long periods. Sugarcane was particularly thirsty. We are told that in the Nile Valley, sugar crops required not only the waters of the inundation but an additional twenty-eight waterings by waterwheel.[8] Spanish farmers irrigated sugar every four to eight days. Other crops, like bananas, cotton, and mangoes, required lengthy waterings. Many crops gave much better yields if watered at a particular time. At the same time, many farmers aimed for year-round cultivation, planting winter and then summer crops in the same fields and doing away with the long fallow periods that were typical of earlier farming practice. Out of sheer necessity, water management was an essential, if inconspicuous, prop of the Islamic world.

Islam inherited domains where irrigation had a very long history indeed, most of it at the village level, but some of it once organized in highly centralized, near-industrial-scale irrigation networks, like those of the Sassanians. Large-scale irrigation had flourished early in the Common Era over much of the Near East, but was in serious trouble during the centuries before the Prophet. Population decline (culminating in the great plague of the sixth century), government neglect, overtaxation, and chronic warfare all played their destructive parts. Canals and aqueducts silted up or were neglected. A major flood in 627 wrought catastrophic damage to embankments and dams along the Euphrates and the Tigris. Thus, Islam also inherited great waterworks in chronic disarray. They brought inadequate irrigation water to too little land, and mostly only at times of flood and rain, during the winter months, not the summer ones. Both the limited technology and the narrow range of gravity devices were inadequate to the task at hand.

Change came with the consolidation of the Islamic state and with a gradual realization that the growth of any town or city depended on the parallel development of both food and water supplies, as well as on increased private and state profits from the land. A new interest in water management saw the repair of old irrigation works and the construction of new ones, as well as the use of a wide range of long-established methods for lifting, storing, and transporting water. A profusion of irrigation technology came into use throughout the Islamic world, most of it old. Rulers and private landowners alike commissioned embankments and trenches on hillsides to capture runoff from gorges and wadis. Villages and towns tapped springs, dug many types of wells, and used ancient *falaj* technology to tap groundwater. Long brick- or stone-lined canals and earthen ditches carried water long distances; so did tunnels carved out of solid rock. Water engineers used lead and clay pipes in some areas and built strategically placed dams to create reservoirs. They developed new forms of cisterns and innovative well-digging methods. But none of these devices solved the fundamental problem of lifting water more than short distances. Islamic hydrologists relied on long-established lifting technologies that could only transport minuscule quantities of water compared with what was needed. The *dalw* was a bucket pulled by a rope passing through a pulley. Farmers and gardeners used pivoted troughs; the *shaduf*, with its counterweight, commonplace in Mesopotamia and along the Nile; and the Archimedes screw, as well as *norias* (waterwheels with revolving pots or boxes). All of these were powered by human labor and sometimes by animals or, in the case of *norias*, water.

What made the difference was the genius of Islamic water engineers in applying such simple technologies on a much larger scale and in combinations that worked well together in new environments. The classic example is the *falaj* technology for tapping groundwater, originally developed in Iran as the *qanat*. *Qanats* were used on a limited scale in Egypt in pre-Islamic times, but their use spread rapidly across North Africa and into al-Andalus soon after the conquest. They became especially important in the arid interior of the Maghreb. Various forms of the *noria* also came into widespread use throughout the Islamic world. These innovations in themselves were effective locally, but their full potential came from integrating them into parts of complex water-management systems to supply cities, towns, large irrigation systems, and royal gardens. Through the use of ingenious combinations of channels, storage facilities, and lifting devices, land could be irrigated cheaply and more extensively than ever before, bringing water to landscapes that had hitherto been parched.

Over the centuries, the Islamic world became a patchwork of irrigated lands. Some areas required heavy watering throughout the year, the fields being under intensive, nearly continuous cultivation. Other lands relied on flash floods and runoff stored in small dams or cisterns that was used sparingly. In many areas, it's no exaggeration to claim that there was hardly a stream, spring, river, or lake that was not fully exploited for its water supplies for domestic, urban, or agricultural use. In some areas, enormous tracts of cultivated and irrigated land carpeted valleys and plains. In the Sawad of Iraq, irrigated acreage subject to the land tax amounted to nearly 19,300 square miles (50,000 square kilometers), nearly all the cultivable land. In al-Andalus's Guadalquivir Valley, thousands of *aflaj* and *norias* watered irrigated lands. One of the latter still stands by the Guadalquivir River, towering over the landscape, for all the world like a "giant, modern erector set."[9] Canals first dug by the Romans now extended in every direction. Such was the prosperity of al-Andalus that the Umayyad caliphate of the West became the conduit through which much of the philosophy and science of classical times, augmented and translated by Islamic scholars, flowed into the Western world. None of this would have been possible without a genius for water management.

Genius there may have been, but it was the genius of hard-won experience, not necessarily of efficient governance. Of course, the fortunes of irrigated lands changed constantly. Rivers changed course, rulers failed to invest in maintenance, inept administration decimated crop yields, wars destroyed irrigation systems. But throughout the Islamic world, there was enough flexibility in water management to ride through low-rainfall years, and even sometimes to expand irrigated lands during such periods. The extent and quality of gravity-driven irrigation works

Figure 15.3 *Waterwheel on the Guadalquivir River. (Ric Ergenbright/ Corbis)*

achieved unprecedented levels of efficiency, even elegance, supporting not only familiar cereal staples but also all manner of new, exotic crops that now flourished in unfamiliar, usually hostile environments.

Over the centuries, the limits of cultivation ebbed and flowed continuously, against a background of steadily increasing populations, especially outside cities and towns. For instance, in Turkistan, new irrigation works drew settlers into lands lying along the tributaries and lower reaches of the Oxus River and into the hinterlands of the great central Asian oases. Farmers pushed settlement into the margins of the Syrian Desert, while drought-resistant sorghum and hard wheat enabled farmers to flourish in the Maghreb, on the edges of the Sahara Desert. Rural population densities rose dramatically. Egypt's Fayum Depression supported 360 villages, said to have grown enough food to supply all of Egypt for a day. There were allegedly 12,000 villages the length of the

Guadalquivir River, in al-Andalus. Along the Tigris River, settlement was so continuous that it was said that before dawn crowing cocks answered each other from housetop to housetop from Basra to Baghdad. The density of rural population, in these places and elsewhere, was a reflection of both the need for more food and the availability of enough people to work the land and maintain irrigation works. By the eleventh century, increased agricultural production also supported rapidly growing cities in every part of the Islamic world, each a center of gravity, as it were, for farmers growing crops to feed city dwellers. The irrigated fields to do this covered increasingly large areas, for the populations of major cities were enormous, far larger than those of any European cities at the time. Baghdad housed between five hundred thousand and one million people; Cairo had at least half a million inhabitants at the time of the Black Death, in 1343 C.E.; even Damascus had one hundred thousand people on the eve of the plague, when the city was past its peak. Córdoba, in al-Andalus, may have had a population of between five hundred thousand and one million people.

The culmination of Islamic artistry with water—it was nothing less—came in the lushness, scents, and shade of gardens, mentioned at the beginning of the chapter. Here, fountains bubbled musically, and inlaid floor tiles displayed intricate patterns; trees provided serenity and shade; palm fronds rustled gently overhead. Many Islamic cities had enormous gardens. Basra, on the Persian Gulf, enjoyed mile after mile of gardens, orchards, and crisscrossing canals. Damascus was said to have 110,000 fruit trees. Al-Fustat, in Egypt, boasted magnificent private gardens. In al-Andalus, Sevilla, Córdoba, and Valencia were veritable oceans of pleasure gardens. The resplendence of Islamic gardens culminated in those built by a ninth-century ruler of Egypt. His gardeners coated the trunks of palm trees with gold sheet. Behind the gold, pipes sprayed water through different openings into nearby pools.

Islamic gardens owed their inspiration to Persian designs, which, in turn, probably went back even further into the past. We know, for example, that the Assyrian monarch Assurnasirpal II devoted much water to his orchards in Nimrud. Many Islamic rulers were enthusiastic plant collectors. The first Umayyad emir of Spain, Abd al-Rahman, sent collectors as far afield as Syria to gather new plants and seeds, among them

Figure 15.4 *El Partal in the Alhambra, Granada, Spain. (Kevlinjay/ iStockphoto)*

the date palm. By the tenth century, the royal gardens at Córdoba seem to have become a botanical garden, complete with experimental plots for exotic plants from every corner of the Islamic world. In Tabriz, in what is now Iran, at the other end of the Islamic world, rulers imported rare fruit trees from China, India, Malaysia, and central Asia. These were no casual hobbies on the parts of their owners. Many gardens came under the supervision of leading botanists. Two leading agronomists, Ibn Bassal and Ibn Wafid, supervised the royal gardens at Toledo. Their agricultural manuals give details of new plants introduced into Spain. Gardens like these linked botanists and others over wide areas and did much to diffuse useful plant forms, as well as the expertise in irrigation and other matters needed to grow them, into new areas. Centuries were to pass before Europe possessed similar gardens, where water management, science, and gardening as an art came together.

THANKS TO RECENT research, we now know much about water management in al-Andalus, which provides a partial view of wider Islamic practices.[10] Here, as elsewhere, hydrotechnology was, for the most part, village based, but the same simple methods came into use for cities, palaces, and towns. Al-Andalus's village-based irrigation methods owed much to migrating Arab and Berber groups, who organized their water management using kin ties, as in other regions we have seen. The great expansion of urban centers saw the emergence of *huertas*. These irrigated lands outside cities relied on marshlands and old river channels. They looked like a comb imposed on the landscape, which now became a network of canals for allocating water. City layout depended to some extent on the canals needed to supply water throughout the urban area for all kinds of purposes. In Palma de Mallorca, for example, the main water canal was fed by a spring 4.3 miles (7 kilometers) from the city. Branch canals diverged from the main artery as it passed through the city. Mosques and other major users lay at the end of larger branch canals. Even the city streets and walls followed the dictates of the urban water channels.

Here, as elsewhere in the Islamic world, everyone depended on proven technologies. Diversion dams and cisterns were the staples of village and fortress water systems. At least twenty-eight vaulted cisterns provided water to Granada, fourteen of them associated with mosques. Wealthy house owners used cisterns sunk below or lying alongside an ornamental pool. Rural cisterns collected rainwater for cattle. *Aflaj* were staples of Andalusian water systems, those on the island of Mallorca often being relatively shallow and short, rarely going below 16.4 feet (5 meters) and using open channels rather than tunnels. Such installations did not require the services of specialist builders. *Norias* were ubiquitous, mostly driven by current and used to raise water from rivers into irrigation canals or to pass water from major to minor channels. Most were donkey powered, the beast being hitched to a horizontal wheel that drove an endless chain of pots. The water passed into a small tank, which regulated the water flow into the fields. Horizontal water mills with often-tall penstocks were virtually identical to those found in Sassanian irrigation systems at the other end of the Mediterranean. Thirteenth-century Valencia boasted at least 385 water mills, 25 of them of large size.[11]

At its pinnacle, Islamic agriculture and water management probably accomplished virtually everything that could be done with existing technology. Virtually all water supplies were in use; rural settlement was so dense that the frontiers of cultivable land had been pushed to extremes in many areas. Inevitably, the upper limits to growth were tested, as populations rose and the limitations of water-harvesting technology became barriers to expansion. The momentum of Islamic agriculture slowed, plagued in part by too-rapid expansion and the familiar problems of overcropped land, rising salinization, and soil erosion, caused by using marginal lands and excessive forest clearance. Islamic agriculture and water management may have been flexible, but both were also fragile. Self-sufficiency was an elusive goal, because of the intensive labor required to build and maintain any form of irrigation works. Preventing excessive runoff and husbanding groundwater were time-consuming, difficult tasks, especially in a world riven by constant violence and warfare. For large-scale, highly productive farming landscapes to survive, the government had to be centralized, concerned about its food supplies, and capable of protecting farmers. The problem was compounded by the patchy nature of settlement in arid lands, where farming and herding took hold in better-watered areas, often with extensive uninhabited zones between them.

The viability of agriculture in semiarid environments depended on water supplies, and these depended in turn on even minor climatic fluctuations. A difference of even a few inches of rainfall annually could extend the margins of potentially cultivable land a significant distance. More important, however, were the effects of droughts, especially prolonged ones, such as were commonplace during the Medieval Warm Period, particularly during the eleventh through thirteenth centuries. The climatic evidence, much of it from lake cores and tree rings, is still incomplete, but there are hints that dry spells were widespread over much of the Mediterranean world during medieval times. That these droughts had a serious effect on Islamic agriculture and water management seems unquestionable, even if the details elude us.

From the eleventh century onward, the decline in Islamic agriculture becomes more evident, as political disruptions and invasions unsettled many parts of the Islamic world. In the West, these incursions culminated in the implosion of al-Andalus in the face of Christian armies. Everywhere conquerors arrived, irrigation works lay in ruins and dams were neglected, intensive agriculture, with its sophisticated water management, being alien to people who cultivated cereal crops in better-watered environments or who practiced much-less-intensive agriculture. Changes in landownership and taxation compounded the decline. In places where European farmers displaced Muslim peasants, new styles of mixed farming on large estates came into prominence, with less-intensive crop rotation. Even on irrigated lands, wheat, barley, and legumes, the staples of the conquerors and their armies, which thrived better in warmer environments than north of the Pyrenees, replaced crops like rice, grown widely in Islamic Spain.

Ironically, Islamic science, with its sophisticated agronomy and knowledge of soils, winds, and water, had fueled centuries of agricultural expansion in every kind of environment imaginable. But there were limits to such science, based as it was on description and cataloging, rather than on seeking to understand how things worked. Had Islamic scientists probed deeper into technology, into such machinery as pumps, they might have anticipated many of the discoveries of the Industrial Revolution. Too many factors conspired against progress in either agriculture or water management—inept government, changing patterns of world trade, a greater prevalence of prolonged droughts, and environmental degradation, among others. But the erosion of earlier achievements should not detract from the brilliant feats of Islamic farmers, gardeners, and water engineers, who deployed the simple harvesting technologies of millennia to the limit, and almost beyond. In some places, like parts of Yemen, you can still see traditional water management in action, continuing an ancient practice of sustainability that is only now giving way to the diesel pump and all the problems of finite groundwater that go with it.

CHAPTER 16

"Lifting Power . . . More Certain than That of a Hundred Men"

"BY MEANS OF A WINDING CHANNEL cut through the middle of the valley . . . by the hard work of the brethren, the Aube sends half of its waters into the monastery, as though to greet the monks and apologize for not having come in its entirety, for want of a bed wide enough to carry its full flow . . . As much of the stream as this [fronting] wall [under which it must flow], acting as gatekeeper, allows in by the sluice gates hurls itself initially with swirling force against the mill, where its ever-increasing turbulence, harnessed first to the weight of the millstones, and next to the fine-meshed sieve, grinds the grain and then separates the flour from the bran." The twelfth-century C.E. Cistercian abbot Bernard of Clairvaux waxed lyrical over monkly harvesting of the River Aube. Nothing was left to chance. From the mill, the water flowed on to provide the community's drinking water, then served the fuller's and tanner's workshops and other facilities before carrying off waste.[1] But as the abbot knew full well, "the hard work of the brethren" and gravity were the only things making it possible for water to be harvested for home, hearth, and field.

Animal and human muscle, waterpower and wind power teamed with gravity—these were the engines of Roman and Islamic water management and the only hydrological technology available anywhere, even in temperate Europe, where water supplies were rarely a problem. If anything, an overabundance of rainfall and flooding presented the biggest challenges to farmer and lord alike in Europe. Between 1315 and 1321, for example, a series of unseasonably cold and wet summers ruined harvests over much of northern Europe. Chronicler Bernardus Guidonis wrote, "Exceedingly great rains descended from the heavens, and they made huge and deep mud-pools on the land."[2] As many as 1.5 million people perished from famine and related diseases during these years, when there was also "a great dying of beasts."

The Romans passed on a legacy of water management that endured even after the empire collapsed. As cities imploded and shrank, their Roman-inspired infrastructures declined and fell into disrepair, despite sporadic efforts to repair aqueducts. Most Roman water systems in Britain did not survive the end of the occupation; some aqueducts in France may have endured for a few centuries. However, the great Roman sewers of Pavia were still in use throughout the Middle Ages. How much hydrological knowledge survived the collapse of Rome in the West is a matter of vigorous academic debate. But there is no question that medieval gravity-flow water systems owed much to the engineering achievements of the classical world. Low-pressure water systems using channels and pipes worked just as well in temperate climates, if not better because there was more water.

Medieval Europe was a predominantly rural world of subsistence

Figure 16.1 *Roman grain grinding on an industrial scale. A staircase of sixteen waterwheels in two rows powered by aqueduct at Barbegal, near Arles, France, ground enough flour for the entire town of 12,500 people. (After A. Trevor Hodge,* Roman Aqueducts and Water Supply. *Copyright © 2002 by Duckworth & Co, London)*

farmers with basic water needs. Most villages and small towns obtained
their supplies from traditional sources: lakes, rivers, springs, and wells.
Royal palaces and dwellings of the nobility sometimes boasted simple
pipes and plumbing, but the first systematic attempts at water management centered on religious houses, the most likely places for at least
some knowledge of hydrology to have survived. The more stable social
environment of the High Middle Ages witnessed a remarkable flowering of monastic communities throughout Europe. Monasteries were
self-sustaining entities, which meant that they had to lie near reliable
water sources such as rivers and springs. The wealthier houses invested
in carefully designed waterworks, based on both in-house expertise and
information gleaned from other religious communities.

The earliest complex water systems that rivaled Islamic and Roman
installations appeared during the eleventh century, many of them in
what is now Germany, where piped water systems graced monasteries
and palaces. Pipes also carried water to religious houses like Chartres
and Cluny in France. Some of the earliest monastery pipe systems served
convents like Abbess Theophanu's in Essen, Germany, which had lead
pipes serving an atrium fountain and the cloister. Archaeological excavations at twelfth-century Norton Priory, in northeastern England, have
revealed a network of open drainage ditches and a hollowed-tree-trunk
drain beneath the latrine block. This primitive water-management system served the priory's first, temporary timber structures. By the end of
the century, the community was expanding and had raised enough resources to erect permanent stone buildings. The builders incorporated
spring-fed lead intake pipes into the priory, and a masonry great drain
fed from the moat to carry away waste. Some monasteries also maintained
fountains outside their precincts for the general public and even allowed
access inside. In the mid-fifteenth century, Bishop Alnwick of Lincoln
complained of Daventry Priory, in central England, that local women
"have general resort to the kitchen and to the washing places in the cloister, where they get up on the edge to fill their pots at the washing places,
and so they befoul the same edge with their feet."[3]

As monastic foundations proliferated, trade and cities also expanded
rapidly, creating demands for clean water. Municipal authorities began
to regulate, albeit loosely, public fountains and the water supplies avail-

Figure 16.2 *Map showing locations in chapter 16.*

able to keep them flowing. These were some of the earliest examples of municipal laws and infrastructure planning for water supplies, enacted partly for thirsty urban industries like tanning and wool processing. Tanners soaked dirty, stiff hides in large water vats, then pounded and scraped them before soaking them in urine. They softened the tanned hides by soaking them once again in freshwater. Wool merchants soaked dirty, greasy wool straight from shorn animals in water to prepare it for spinning. Urban water supplies assumed great complexity in later medieval times. Paris, and perhaps Siena, instituted a public water system in the 1190s. London acquired springs for its Great Conduit in 1237. All these efforts faltered in the face of inexorable urban growth. Sheer numbers of people gradually stressed the capacity of traditional supply sources; more and more urban waste endangered both groundwater and public health.

BY THE STANDARDS of today, Europe's medieval cities were tiny. Almost everyone lived in the countryside, where castles, monasteries, and nobles' dwellings used water for defense, as well as for pleasure and aesthetic satisfaction. The canals that flowed through Clairvaux Abbey provided water for drinking and washing, as well as for artisans, but also "served a dual purpose of nurturing fish and irrigating vegetables."[4] Some moats surrounded orchards and herb gardens, like one at Peterborough Abbey, in eastern England, which encompassed 2 acres (0.8 hectares) of herb gardens and pear trees. Ornamental gardens with water-lily-strewn ponds and rippling streams were as favored among Europe's nobility as they were in the Islamic world. Nearly every monastery and noble dwelling had its own fishponds, which served both ornamental and, more often, practical purposes.[5]

Fish played a vital role in medieval diet, prized for its spiritual purity, unlike meat, which was thought to encourage carnal thoughts. Meat was proscribed for holy days and Lent, part of a diet that atoned for Christ's suffering on the Cross.[6] By the thirteenth century, nearly half the days of the year were technically holy days, so the demand for fish was insatiable. River fish such as bream, perch, and pike provided a welcome fresh alternative to the ubiquitous dried eels, salted herring, and cod. A large breeding pond, or vivarium, sometimes a small lake, nurtured growing fish, kept fresh by slow-flowing current. The ponds were drained about every five years, the fish being transferred to smaller storage ponds where they could be readily caught from the banks. Monastic ponds never provided enough fish to feed an entire monastic community. At least 2 acres (0.8 hectares) of water was needed to provide daily fish for a single monk. Vivaria were symbols of prestige, sometimes constructed at great cost and at great inconvenience to neighbors, the fish being used only on special feast days.

Fish farming became big business in medieval Europe. The first large-scale fish farms developed between the Loire and Rhône rivers around the eleventh century, based on local species like bream or pike. Such enterprises really took off when carp, which spawn in very shallow, muddy water, appeared across Europe, arriving from their Danube home via tributaries of the Rhine by the twelfth century. Carp farming spread rapidly through France and into the Low Countries over the next two centuries. By the 1340s, the Cistercian monks of Heilsbronn Abbey, near Nuremberg, managed a carp-farming enterprise that extended more than 18.6 miles (30 kilometers) along rivers and low-lying meadows. Czech farming operations involved elaborate sluices and valve systems that maintained ponds covering 247 acres (100 hectares) or more. Some of the most elaborate farming operations involved staircases of ponds created with clay-and-loam dams with stone or compressed-clay cores. The Barons Rozmberk of Trebon, in southern Bohemia, maintained elaborate pond complexes in marshy terrain along a tributary of the Elbe River that supplied carp to Prague and other cities. Some 154 square miles (400 square kilometers) of fishponds still thrive around Trebon today.

The acreage devoted to fish breeding was staggering. A huge swath of Europe from the Loire and Poitou to central Poland supported fish farming wherever the terrain and water supplies permitted. There were twenty-five thousand fishponds in Bohemia alone and more than 98,800 acres (40,000 hectares) under ponds in central France, all in areas too far from the ocean for carts or barges to bring in fresh sea fish. The profits could be enormous, but the labor involved was astounding. You can leave a field in fallow and the soils will eventually recover. But a fishpond suffers from neglect within months—breached dams, encroaching vegetation, and silting were only a few of the problems faced by pond owners. Ponds soon fill with sediment as the natural drainage reasserts itself. For this reason, aquaculture was not self-sustaining. When improved salting methods brought large stocks of cod and herring to refectory tables, fish farming went into decline, except in eastern Europe. Virtually all of France's once-bustling fish farms are now dry land.

A THOUSAND YEARS ago, almost all of Europe depended heavily on the backbreaking work of farmers and their animals. Fortunately, the Romans left behind an important technological legacy: the water-powered grain mill. Centuries of experience with waterpower produced innovations that were to change water management everywhere. Terry Reynolds, a leading expert on water mills, once remarked that "if there was a single key element distinguishing western European technology from the technologies of Islam, Byzantium, India, and even China after around 1200, it was the West's extensive commitment to and use of water power."[7]

Norias were well established in the eastern Mediterranean world by the first century B.C.E. These were vertical, so-called undershot wheels, rotated by water striking paddles or blades at the bottom of the wheel. (The term "undershot" refers to water striking the wheel at the base. The turning wheel, powered by the stream, scooped up the water.) At some point, an anonymous innovator attached gearing to the wheel that enabled it to turn a large millstone. Such devices moved the poet Antipater of Thessalonica to lyrical prose: "Rest your mill-turning hands, maidens who grind! Sleep on even when the cock's crow announces dawn, for Demeter has assigned to the water nymphs the chores your hands performed. They leap against the edge of the wheel, making the axle spin, which with its revolving cog turns the heavy pair of porous millstones from Nisyris."[8]

Here, the Roman legacy is unmistakable. We've already met the industrious Marcus Vitruvius Pollio, who devoted the last volume of his *De Architectura*, of about 25 B.C.E., to machines. He described five types of waterwheels, among them the undershot design and flour grinders, the latter the so-called Vitruvian mill, with two large gears that meshed at ninety degrees, the size of the gears governing the speed of the millstone. Romans ground flour on an industrial scale, using aqueducts, a controllable water source, to turn their grinders. Few such mills survive, but an impression of an undershot waterwheel preserved in the ash from Mount Vesuvius at Pompeii was 6.5 feet (2 meters) across, with eighteen radial spokes. This wheel would have been capable of sustained high-speed operation. Throughout the Roman Empire, waterwheels provided a basic service to communities of every size, and also to military encampments. In 575 C.E., Gregory of Tours described how the stream before the gates of Dijon, in France, "turns mill-wheels with marvelous speed."[9]

Descendants of the Vitruvian mill ground grain during the European Middle Ages. Monastery communities, concerned as they were with self-sustainability, located themselves in places where water mills with grinding wheels, perhaps attached to pounders, would work effectively. Small innovations improved the gearing, but were mainly directed toward improving the flow of water to the wheel. Generations of anonymous carpenters and millers maintained a water technology that survived the constant upheavals and internecine warfare of the first millennium, but their efforts go unrecorded by history, except in scattered estate and monastery records. Water mills were commonplace. In 1066, William the Conqueror's *Domesday Book*, an inventory of his new domains, listed as many as 6,082 English water mills. These may have met up to a third of England's energy requirements.

By no means were all of these mills of vertical design. Many used small horizontal wheels that had the advantage of not requiring gearing. Paddles in the stream below the grinding wheel turned a vertical axle and the grinder. These eventually disappeared, as lords, or organizations of merchants, invested in more powerful vertical mills, not only for their own use but also as a way of making money by grinding grain for others.

Bread was Europe's dietary staple, its production filling the time of tens of thousands of people, from plowing to harvest to grinding grain and baking loaves. Small wonder that the water mill assumed enormous importance in medieval society, commemorated in texts and the art of the time, for grain was the source of both bodily and spiritual nutrition. The New Testament records a meal that assumed profound symbolic meaning: "And as they were eating, Jesus took bread and blessed it, and brake it, and gave it to the disciples, and said: 'Take, eat, this is my body.' "[10] Bread and wine were body and blood, a diet that was a key to deliverance. Early religious mill images portray a wheel turned by the twelve apostles, or a waterwheel turned by the four rivers of paradise. The mill transformed evangelical prophecy into Christ's physical body, crushing grain to create meal and wafers just as the Savior was crushed

Figure 16.3 *Horizontal water mill. These cheap, relatively inefficient mills had no gearing and required a fairly small volume of water. (Terry S. Reynolds,* Stronger than a Hundred Men. *Copyright © 2002 by the Johns Hopkins University Press. Reproduced with the permission of the Johns Hopkins University Press)*

by his suffering and delivered salvation. Such depictions metaphorically created the body of Christ, improving the grain and, at the same time, creating a virtuous substance.

Blessed with generally abundant water and an amplitude of flowing streams, Europe never completely abandoned Roman milling technology. In addition, famines and plague contributed to a dramatic fall in European population that lasted until the fourteenth century. Manpower shortages on the land created strong incentives for adopting labor-saving devices like water mills. Water mill construction accelerated rapidly from the mid-twelfth to mid-thirteenth centuries, a function of both increased demand and economic reality. Waterwheels came into their own in medieval Europe, and continuing demand prompted constant efforts to improve their power and productivity. In 1540, the Italian mining engineer Vannoccio Biringuccio wrote that "the lifting power of a [water]wheel is much stronger and more certain than that of a hundred men."[11] He was probably correct, for even a small overshot water mill with just two horsepower could free between thirty and sixty people, most likely women, from grinding grain.

Vertical undershot wheels could, theoretically, operate anywhere on a riverbank, but their efficiency rose markedly with a swift flow, directed through an artificially constructed millrace, a narrow waterway that produced a steady, rapid current, above 4.9 feet (1.5 meters) per second.[12] Building a millrace required sometimes quite elaborate construction that could involve dams, reservoirs, retaining walls, even aqueducts. But such devices were effective in combating a perennial problem: the changing flow rates of rivers at different times of the year. Waterwheels could be built into bridges, but the flow problem still remained. A race and its dam sluices controlled the water flow and, if set up at a higher elevation, could not only add velocity to the flow but also divert water to a mill situated at the most convenient location, sometimes some distance away from the water source.

Water mills were among the most prized possessions of the elite, who earned substantial revenues from operating them. As time went on, control and ownership of water mills passed from small communities and towns into the hands of the aristocracy and monasteries. Judging from statistics from thirty religious houses in England, in 1086 the Benedictines may have owned as many as two thousand of the six thousand or so water mills in the country. The Benedictines were the richest order and the monks most involved in mill construction. During the thirteenth and fourteenth centuries, between 6 and 10 percent of a monastery's income could come from milling.

Throughout Europe, the lay sector was well aware of the profitability of water mills, but much depended on the costs of building and maintaining one. The cost of constructing an entire water mill complex, complete with millponds, sluices, and so on, could be as high as twenty pounds, about enough money to half-fill a small automobile gas tank today, but a considerable sum for the time. Maintenance costs could average about 20 percent of income, but once the low wages of the miller were paid, everything was free and clear. The average profits for a water mill could be as high as 70 to 80 percent of revenue, especially if lordly monopolies were in force. However, many people went to independent mills and were not subject to the often confiscatory charges of lords and monastic houses.

THE OVERSHOT WHEEL, a medieval invention, revolutionized milling. An elevated aqueduct delivered the water to the top of the wheel, where it poured into buckets or other containers on the wheel rim. Not the rate of flow, but the weight of the water hitting these receptacles governed the speed of the wheel. Add a millrace and a dam to the delivery system, and the wheel had real power and dependability. A well-constructed overshot wheel was more than twice as powerful as an undershot, while needing about a quarter of the water. They were more expensive to build, but more than paid for themselves.

As with undershot wheels, we know almost nothing about the overshot mills' early history, except that they were probably rare until the early sixteenth century, perhaps because they were much more expensive to build. The overshots, though, offered advantages in terms of capital conservation, this being measured, in predominantly agrarian economies, in livestock. Horses and mules could turn mills, but they had to be fed; as populations rose dramatically between the tenth and fourteenth centuries and again between the sixteenth and seventeenth, many more animals would have been required—and the forage to feed them. The savings in terms of harvested production would have been enormous.

For centuries, medieval Europeans used waterpower almost exclusively for grinding grain, not only for flour but also for pulverizing barley into malt for another important staple: beer. Beer mills date back to the ninth century in France. By the eleventh century, modified water grinders in the south also processed olive oil by crushing the seeds, and soon such crops as mustard and sugar. Two centuries later, water mills were crushing oak bark into tiny fragments all over Europe. This accelerated the leaching process that produced tannin. They also drove carborundum stones for polishing and sharpening everything from cutlery to swords and axes. Waterpower cut timber of all kinds; by 1590, waterpowered lathes had appeared in some carpentry shops.

Then there were mines. By the mid-fifteenth century, there was a chronic gold and silver shortage in much of western Europe. The Romans had exploited the most accessible silver-ore seams, in southern Germany and central Europe.[13] Silver mines lay in mountainous, well-watered areas, where deep-shaft flooding was a major obstacle and pumps were essential. Roman-vintage water-powered bucket chains and screw pumps simply did not have the lifting power needed to clear passages far underground. Then the water-powered suction pump came along in the mid-fifteenth century, a basically simple cylindrical device that expelled air and drew in water. Overshot wheels operated the piston rods that expelled air, forcing water up to the next level of the deep shaft, where another pump took over. Miners could now reach much deeper, rich seams; the result was a silver-mining boom that increased Europe's silver supplies by at least fivefold, until cheaper silver crossed the Atlantic from the Americas during the sixteenth century.

Mine pumps were an important catalyst that made Europe more dependent on pumps of all kinds than any region in the Byzantine and Islamic world. By the fifteenth century, waterpower was spreading multi-faceted tentacles throughout the late-premodern European economy. For instance, the development of water-powered trip-hammers produced much larger quantities of purified iron at a drastically lower labor cost. (Trip-hammers are raised and allowed to fill by a lever or some other tripping device.) Another fifteenth-century innovation linked water-wheels to bellows, which maintained much higher temperatures in tall, brick smelters, some as much as 29.5 feet (9 meters) high. The higher temperatures inside the furnace allowed the ironmaster to pour molten ore into molds and produced cast iron with a high carbon content. Further refinements, involving the use of water-powered piston bellows and steam, did not come until the mid-eighteenth century, in the early decades of the Industrial Revolution.

The first full-scale industrial application of waterpower outside mine pumping was for fulling woolen cloth, the process whereby heavy fabric of high quality was degreased, cleaned, matted, and felted.[14] Teams of fullers trod on the cloth for three to five days as it lay immersed in a vat of warm water, urine, and fuller's earth (an abstract, opaque clay), working up to 240 days a year and for as many as fourteen hours a day. A single team would produce about thirty to thirty-five cloths a year, with a total length for all of them of about sixty-nine feet (twenty-one meters). Water-powered fulling mills operating trip-hammers needed but one man to operate them and processed a cloth in a day or so. When they first came into use is uncertain, but it may have been as early as the tenth century. By the fifteenth century, a rural-based, water-powered English cloth industry outpaced its rivals in the Low Countries and elsewhere.

Mechanized fulling resulted in a significant productivity jump, perhaps as much as a thirty-five-fold increase. Another innovation involved a water-powered gig mill, developed during the early fifteenth century. This device mechanized the process of napping cloth to finish it by passing rapidly turning metal cylinders across the back and front of the cloth. By the seventeenth century, two men and a boy could operate a gig mill and do the work of eighteen men and six boys, while reducing the labor time from one hundred to twelve hours in a nine to one productivity gain. At the luxury end of the cloth business, waterpower also played a part in silk manufacture, where water-powered throwing machines were producing good-quality silken yarn as early as the twelfth century in Italy. Once again, there were obvious labor savings. One or two operators could do the work of several hundred hand-throwsters. Waterpower of all kinds had achieved a considerable level of sophistication in the European world by 1700, on the eve of the Industrial Revolution.

MEANWHILE, EUROPE'S CITIES grew and grew, becoming veritable mazes of competing water interests. London is a case in point.[15] As many as eighty thousand people lived there by the thirteenth century, and the medieval city was a jigsaw puzzle of different water-management entities. Long strips of land extended back from narrow street frontages, the building fronting on the street while the rear of the property was an open yard, sometimes with outbuildings that served as privies and for other purposes. Some tenements sank their own wells into the underlying river gravels. Most people relied on rainwater collected in cisterns, supplies from public sources, or freshwater streams. They could also purchase water from professional water bearers. Veritable honeycombs of refuse pits lay in domestic yards. Most cess- and garbage pits were relatively shallow, but some of them penetrated into porous gravel beds, where their contents contaminated the groundwater.

As the city became more crowded and land was subdivided even further, rainwater and waste from overflowing cesspits flooded neighboring yards and cellars and rotted timber walls. Disputes between neighbors became so prevalent that the authorities developed a public drainage network, compiled building codes, and passed an Assize of Buildings around 1200 that tried to regulate gutters, eaves, latrine pits, and other sources of contention. Citizens were not allowed to throw water or liquid waste out of their windows, but they could dispose of them down public kennels, open channels running down the middle of the city's streets that drained into the Thames and its tributaries. In some places, house owners built latrines that deposited waste directly into the river below. The pollution from domestic waste obstructed watercourses, this apart from the entrails and decaying bones deposited by butchers from their own wharf on the Fleet River, which flowed into the Thames. Fullers and dyers built their workshops along riverbanks, also contributing to pervasive contamination of the water supply. Many house owners built their own revetments along the Thames, pushing them out into the river, so as to avoid the accumulation of waste along their small length of shoreline. Fortunately for them, the Thames, being a tidal river far upstream of London, was too irregular and fast moving for water mills. Much of London's grain came from mills located along the nontidal reaches of secondary rivers. One such was the River Lea, 4.3 miles (7 kilometers) to the east.

During the thirteenth and fourteenth centuries, both the city authorities and religious houses began drawing freshwater from distant springs by constructing piped water channels. Some such water systems were carefully planned and served the needs of small communities, but London's civic water system was inadequate for the needs of eighty thousand people. The city still relied on wells and polluted rivers, but work on a comprehensive project continued, perhaps inspired by a water network installed at Westminster Palace by Henry III in 1234. The civic system grew piecemeal over the centuries, as progressively more springs and distribution points came online. Some Londoners even developed their own pipelines, which, legally or illegally, drew water from the public supply. In general, London eventually developed a water system that worked, for all its haphazard jumble of ecclesiastical, private, and public water management. The drainage network became more coherent, but the major problem was pollution, caused in part by poor regulation of private waste-disposal practices. London's freshwater conduit system was advanced for its time, and almost unique in Britain, and the conduits produced reasonably pure water, certainly better than that from the nearby rivers. But there was never enough to reduce the dependence on rivers and wells. London's authorities had some success in regulating and supplementing, and even linking, a patchwork of private water schemes, but they could never replace them with a comprehensive, coherent water-management system, something that was beyond medieval hydraulic technology, based as it was on gravity.

When cities were able to acquire new water sources and had the money, they would expand existing ones or build new ones. In England, the suppression of the monasteries by Henry VIII from 1536 to 1540 was a windfall for cities. Bristol's parishes acquired monastic conduits. The king granted Cambridge's Franciscan conduit to Trinity College. It still feeds the college fountain. Some urban systems expanded too far and had to be scaled back. In Paris, so many private pipes came into use that the water supply to public fountains became a trickle. Charles VI ordered the pipes destroyed in 1392—except those owned by himself and his family. The situation never improved. By the sixteenth century, a new boom in private pipes again stressed the water supply. Inevitably, the public demanded more private pipes and more fountains. Some backers of new schemes advocated supplying larger quantities of less pure water from urban rivers, thereby reducing the distance that freshwater had to be transported. After medieval times, long, river-fed canals provided an admittedly expensive solution, like London's New River waterworks. At the same time, combinations of waterwheels and simple pumps originally developed for draining mines raised water from rivers, which was stored in water towers, then delivered throughout the city under pressure through wooden, rather than expensive clay or lead, pipes. Such pumping devices appear to have originated in Germany and Switzerland during the fourteenth and fifteenth centuries.

Figure 16.4 *One of George Sorocold's waterwheels built at the London Bridge Water Works, circa 1700. With a diameter of 20 feet (6.1 meters), it drove eight pumps using crankshafts and rocking beams, raising more than 2,500 tons (2,268 tonnes) of water 120 feet (36.6 meters) a day. The axle sat on pivoted levers so that it could be raised or lowered as the rise level changed.*

In 1581, one Peter Murices from the Low Countries built a water engine in the first arch of London Bridge that worked force pumps. (Force pumps are piston based, with two chambers. As one sucks in water, the other expels it.) Murices delivered river water to houses by pipeline in the eastern part of the city—for a fee. Such private systems proliferated. In 1701, George Sorocold built an improved water engine at London Bridge. His design was so successful that many provincial towns installed his device. These and other such schemes were, however, aimed at paying customers, not at the general public. Providing public water service became a lower priority. Private companies became so pervasive in Britain that by 1846 only 10 out of about 190 municipal councils controlled their city's waterworks. As early as 1752, two primitive Newcomen beam steam engines pumped Thames water into private reservoirs in Green and Hyde parks, in what is now Central London.[16]

Figure 16.5 *Two Newcomen beam steam engines pump water from a canal to reservoirs at Green Park and Hyde Park, in the heart of London, in 1752. Their owner, the Chelsea Waterworks Company, was founded in 1723 and continued to provide water to many Central London locations until 1902.*

The issue in London was not merely insufficient supplies, but also serious contamination of the water that was available. Untreated water may have helped solve the problem of abundance, but consumers were at increased risk of exposure to waterborne diseases as urban populations increased rapidly and pollution intensified with the dawning of the Industrial Revolution. The situation had reached critical levels by the mid-nineteenth century, when science demonstrated how contaminated water transmitted disease. The great London cholera epidemics of 1848–49 and 1853–54 led to a series of parliamentary reports concerned with sanitation and water purity.[17] The tangle of private supplies and rampant water contamination in an age before germ theory had become a potential public health time bomb. A Board of Health report of 1850 criticized the quality of water provided by private companies and called for a publicly administered supply. Portentously, the board remarked, "It has from earliest times been recognized as the duty of government to take cognizance of running waters." Then in 1869, a Royal Commission on Water Supplies called for municipal waterworks that provided pure water to all. The Victorian reformers invoked the medieval past and praised many ancient practices as the modern era of London water began. However, they were not aware of just how backward European water practices were compared with those of Greece and Rome two thousand years earlier, and those of Islam to the south of the Pyrenees in more recent times. At the beginning of the twentieth century, London's water companies were nationalized and formed into a unified entity, now called Thames Water, which supervises an aging system of iron pipes, gradually being replaced with more durable plastic. Most of London's water still comes from the Rivers Lea and Thames, as well as from underground sources.

MID-EIGHTEENTH CENTURY BRITAIN was still predominantly agrarian, with most people living in or close to poverty, even if change was afoot. Animals, humans, water, and wind still powered the economy; gravity still governed most water supplies. Then, suddenly, profound changes in the cotton industry challenged the tyranny of gravity. Traditionally, Britain's cotton industry had been unable to compete with Indian calicos and muslins in either quality or price.[18] Then, in the 1750s, demand skyrocketed at a time of difficulties with Indian supplies. New ideas were in the air. James Hargreaves's spinning jenny, a multispool spinning wheel, invented around 1764, worked with 8 spindles and changed the economic equation. By 1800, the larger jennies operated 100 to 120 spindles.

Spinning jennies boosted productivity dramatically, but it was Richard Arkwright's water frame, patented in 1762, that really laid the basis for the textile revolution, for it created pure cotton products for the first time. A waterwheel drove a number of spinning frames with greater power than a human could. Unlike the jenny, which could be operated in a cottage, this was exclusively a factory machine, at first designed to be horse operated, then later to be powered by water and steam. In 1771, Arkwright installed a water frame in his cotton mill on the River Darent, in Derbyshire, one of the first factories designed around machinery and hours of work rather than daylight hours. Productivity jumped. A few years later, Samuel Crompton's spinning mule, patented in 1779, combined the jenny and the water frame, procuring smoother and finer yarn. Within a few years, crippling limitations on cotton output vanished and large-scale factory industry became practicable. By 1812, a single spinner could produce as much fabric in a fixed time as two hundred spinners had in earlier times.

Within little more than twenty-five years, the cotton industry became one of the most important industries in Britain and outstripped wool production. At first, this was a localized industry, concentrated in the northwest, in Lancashire, where labor was abundant, the water was of good quality, and fast-running rivers and streams were close at hand. Waterwheels also powered the iron industry, although they were soon replaced by steam pumps, which also pumped water from deep coal and iron mines.

Water may have helped spur the Industrial Revolution, but two other innovations really precipitated a continual process of industrialization.[19] One was steam power, the other the development of a paddling and rolling process that enabled the production of cheap, malleable iron. The Scottish engineer James Watt built his first commercially viable steam engine in 1776, which provided immediate access to much deeper coal and iron seams and allowed operators to pump water from considerable depths in much larger quantities than ever before. For the first time, people living on the earth's surface could tap water supplies that were inaccessible with smaller pumps or through gravity harvesting. Seasonal fluctuations in water flow did not interfere with steam power. No longer did factories have to locate in remote rural areas where water was available but labor was in short supply. Waterwheels were the last flicker of the preindustrial world, although, of course, they still remained in widespread use for grinding village grain for many generations, right into the twentieth century.

Waterwheels vanished slowly in the face of steam power. The numbers tell the tale. In 1839, there were 3,051 steam engines and 2,230 waterwheels operating in Britain's textile factories. An engineer named John Smeaton made important changes to overshot-waterwheel design, introducing iron axles and gears. He claimed that he could double the power of a wheel with the same amount of water and even used steam engines to pump water to turn waterwheels. Another engineer, T. C. Hews, built waterwheels all over Britain, many of them with gearing around the inner circumference. Larger and more efficient waterwheels included a giant iron wheel 72.5 feet (22 meters) in diameter that pumped water from lead mines on the Isle of Man. This machine weighed 200 tons (181 tonnes) and developed about 200 horsepower (149 kilowatts).

By now, the scale of water affairs had changed profoundly. A seventeenth-century engraving of Germany's Oder Marshes depicts a labyrinth of natural channels and swamps, over fenlands rich in birdlife. A century later, King Frederick the Great gazed proudly over a reclaimed landscape that extended to the horizon. During the eighteenth and nineteenth centuries, Germans drained, diverted, and dammed rivers and lakes, marshes and swamps, in a massive remaking of the landscape that reflected drastically changed attitudes toward the natural world— and toward water. By the mid-nineteenth century, water was an industrial commodity to be pumped, bought and sold, and redistributed in ways that were unimaginable in earlier times. Such commoditization became a matter of legal disputes. In New England, for example, conflicts arose over water rights in local rivers, which provided waterpower for textile mills. Water flow was now so valuable that mill owners downstream sued those upriver, accusing them of diverting water.[20] These law cases were a portent of a future in which humans had finally triumphed over the tyranny of gravity, with lasting implications for our own world.

CHAPTER 17

Mastery?

I LOVE THE SENSUOUS FEEL of freshwater running over my arms after a long, hot day in the sun. A turn of the faucet, and the crusted salty spray on my face vanishes in moments, leaving a wonderful feeling of well-being. Wading in glacial meltwater in New Zealand, a glass of cold water after hours in Egypt's Valley of the Kings, the incredible luxury of a mugful of hot water to shave with after hours in a dusty trench—water has caressed my senses so many times that I feel its special bequest from the natural world. Maybe I'm unusual in these feelings. Or maybe it's that I've spent most of my life in semiarid landscapes where every drop counts, or on passage in small boats for days on end with only tanks and jerry cans to rely on. I refuse to think of water as a commodity like electricity, gravel, or oil. Our blue planet has many gifts to bestow, but none of them approach that of water, truly our lifeblood. Water is life and has been since the very beginning. But today, our reckless consumption of this clear liquid, the elixir of life, has taken us far beyond the point of sustainability.

An endless flow of water links every part of our existence. Our forebears realized this, and never took it for granted. They knew that their lives, their sustenance, well-being, and bodily health, depended on this erratic and completely indifferent substance. Water itself is oblivious to human needs, moving between extremes of flood and drought, a force that is placid one moment, raging in turbulent flood the next. Lakes evaporate, rivers change course without warning, sudden inundations sweep away centuries of irrigation works. For all our efforts to channel and control it, water governs itself and often defies capture. Small wonder that the Romans thought of free-flowing water as a wild animal, only subject to human law when corralled in aqueduct or canal. The capricious moods of water lay in the hands of gods, goddesses, and temperamental water nymphs.

Nearly every ancient society enjoyed close spiritual relationships with water—the Australians with the Dreaming and its signposts to water holes, the Egyptians with the sacred waters of the Nile that brought fertility and symbolic rebirth, the Maya with the dark, primordial waters before the creation. Water was powerful, a matter of life and death, part of a sentient landscape brought to life by gods and goddesses, often in wells, rivers, and lakes. In the Old Testament, the prophet Ezekiel writes of the Holy Spirit: "I will sprinkle clean water . . . A new heart will I give you, and a new spirit I will put within you."[1] The Holy Quran proclaims how water made all living things, a merciful gift from God.

A merciful gift from God—these words resonated in my mind as I dunked my salt-encrusted head under a garden faucet after an afternoon sail yesterday. The water was cool and refreshing, truly a benison. I remembered how medieval Christians considered water the essence of life, dispensed gracefully by great abbeys and monastic houses to dependent communities. Providing clean water was both a charitable and a spiritual act. Then I remembered a recent newspaper article about the cost of water per unit in a community near Los Angeles and realized, not for the first time, how much our attitudes toward water have changed since the Industrial Revolution.

Water was a secular commodity in the more remote past, but usually with underlying spiritual associations. Not, perhaps, for the Sassanians, who devastated Mesopotamia with their enormous revenue-driven irrigation works. But while the Romans took aqueduct water for granted, almost as an entitlement, powerful supernatural beliefs still persisted, especially in the countryside. The secularization of water in Britain coincided with the decline of the monasteries after 1500, when towns and cities strove to provide water supplies to growing urban populations. In the fields and villages, power over water lay in the hands of the local people: farmers, millers, watermen, who maintained well-watered riverside meadows and weirs, and landowners. There was little pressure on their water resources, each person having equal rights to water flowing past his or her land, for watering cattle and fields, for mills and water meadows. All of this required close cooperation between neighbors living in small, stable communities where long- and short-term economic and social ties provided a secure framework for negotiations. For centuries, English villagers living along rivers defined their transactions with one another in large part by water. These kinds of small-scale, locally based, and collaborative relationships are very effective in maintaining self-sustainability.

When the seventeenth-century French philosopher René Descartes proclaimed that humans were "the lords and masters of nature," he reflected a growing sense that Western civilization's destiny was to dominate and tame the earth. Such mastery could take many forms—making "proper use" of the land for cultivation, reclaiming wetlands, straightening rivers, obtaining reliable water supplies from afar. The quest for mastery quickened as intensification of agricultural production and other land use put greater pressure on water resources. Such pressures had operated many times in history, but never with such enormous ambitions as those that surfaced on the eve of the Industrial Revolution. A sense of conquest, of supremacy, replaced respect for water. The poet Oliver Goldsmith rejoiced in his 1784 *History of the Earth, and Animated Nature*, "God has endowed us with abilities to turn this great extend of water to our own advantage. He had made these things, perhaps, for other purposes; but he has given us faculties to convert them to our own . . . Let us boldly affirm, that the earth, and all its wonders, are ours; since we are furnished with powers to force them into our service."[2] By this time, Frederick the Great of Prussia was remaking entire landscapes.

NO ONE COULD accuse Frederick of thinking on a small scale. Three years before he became king in 1740, he declared that "making domain lands cultivable interests me more than murdering people."[3] Frederick's passion for remaking the landscape relied on both human labor and gravity. He commissioned a channel to shorten the Oder River's course by 15 miles (24 kilometers) to improve navigation, increase river flow, and reclaim the huge Oder Marshes. The project depended on picks, shovels, and buckets. Hundreds of fever-ridden workers labored waist deep in mud. Ice broke dikes and caused floods that inundated nearby towns. Finally, success: In 1753, Frederick gazed down on the "smooth and tranquil waters of the canal." "I have conquered a province in peace," he remarked. His undertakings paled beside engineer Johann Gottfried Tulla's shortening of the flood-prone upper Rhine from 220 to 170 miles (354 to 274 kilometers) by massive earth moving, removing over 2,200 islands, and building 160 miles (247 kilometers) of dikes, a project that took most of the nineteenth century. Tulla wrote that humans should shape the river for their own purposes: "In cultivated lands, brooks, rivers and streams as a rule should be canals, and where the water flows should be in the power of the inhabitants."[4]

These confident, often startlingly bold "improvements" came at a time of profound changes in rural and urban life. Inland waterways, deep mines, railroads, and improved highways—the Industrial Revolution conquered time, space, and nature in ways that were unimaginable in earlier centuries, with profound consequences for human relationships with water. In a devout age, sanitary reform—proper water supplies and sewers—became a social responsibility, the "will of God." And for the first time, hydrological engineers could harness steam power to turn powerful turbines in order to extract water from deep aquifers. By the late nineteenth century, pressure water, the exploitation of aquifers, and lengthy aqueducts had ushered in a new era in water management reflecting an industrial age. No longer was water revered. It was merely a commodity to be bought and sold, something taken for granted in a world where city populations numbered in the millions.

THE CHANGEOVER OCCURRED gradually on both sides of the Atlantic, as water assumed great legal value and technology improved. Many New Yorkers drew their water from wells and cisterns and carried it to their homes until well into the nineteenth century. Wooden pipes were still commonplace in East Coast cities, which were convenient for firefighters. They would punch a hole in the pipe, attach their hose, and connect it to a two-man pumper, resealing the hole after the fire was out. But times were changing. In 1820, architect Isaiah Rogers designed the four-story Tremont House hotel in Boston, the first with as many as eight water closets, situated at the rear of the central court on the ground floor. Water for the toilets and the bathrooms in the basement came from a water tank on the roof. Rogers's six-story Astor Hotel in New York City had seventeen rooms, with bathrooms and water closets on the top floor and a storage tank on the roof filled by a steam pump. The city's inadequate cast-iron pipeline system, which supplied public fountains and some houses, broke down during the great fire of 1835, resulting in the development of a pressurized system that drew water from a large reservoir in Croton, forty miles (sixty-four kilometers) north of the city. In New York and elsewhere, water delivery from a distance solved long-term supply problems. Aqueducts became major sources of urban water, notably New York's Catskill Aqueduct system, built between 1907 and 1924. The 163-mile (262-kilometer) aqueduct uses gravity to transport water at a rate of about 4 feet (1.2 meters) a second, through conventional and pressure tunnels, 6 miles (10 kilometers) of steel siphon, and open conduits. About 350 to 400 million gallons (1,324,894 to 1,514,165 cubic meters) of water flow through the aqueduct daily, well below full capacity. Perhaps the most remarkable achievement was that of Chicago during the 1850s and 1860s, when the city built a twin-tunnel system that extended 2 miles (3.2 kilometers) out into Lake Michigan. A massive standpipe 138 feet (42 meters) high equalized pressure throughout the system.

All these infrastructures tapped existing and relatively accessible water supplies replenished by annual rainfall. Such systems were sustainable over long periods of time. But what about the much drier environments of the American West, where urban populations skyrocketed after the gold rush?

Four hundred years after Hohokam society dispersed, Mormon settlers arrived at the Great Salt Lake. Brigham Young himself declared that this was their new homeland, Deseret, a term for the honeybee in the Book of Mormon.[5] Within days, the brethren had a small dam and ditches irrigating a 5-acre (2-hectare) field of potatoes. Their tools were of the simplest design, their dams scraped together from earth and stones, but their willpower was inexhaustible. The pioneers lived in a closely knit society and the church decided where to capture water and how to distribute it. The water from the nearby Wasatch Mountains belonged not to individuals but to everyone, resulting in small-scale communities that were sustainable over many generations. By 1910, small-time farmers in Utah irrigated nearly 1 million acres (405,000 hectares).

Small groups of farmers living in sustainable communities—the idea appealed strongly to John Wesley Powell of Colorado River fame, who spent time among the Mormons and acquired a profound knowledge of the desert West at a time when few Easterners had been there. On the face of it, the obvious thing to do was to leave most of the West unpopulated except for isolated, better-watered enclaves. However, such thinking ran contrary to doctrines of progress, western expansion, and proper use of the land for agriculture. Besides, America would become two lands separated by desert. Only one weapon would suffice to master the arid lands: water. The dream was golden, a glorious vision of fertility and prosperous farms, and a wonderful concept for ambitious politicians. But only one person really understood the potential for western agriculture: Powell.[6]

Powell was director of the United States Geological Survey, an honest man surrounded by corrupt operatives in and outside of the government who were enticing poor settlers to move west onto cheap land, provided they could irrigate it within three years. No proof of irrigation was required. On April 1, 1878, Powell submitted a 195-page document, *A Report on the Lands of the Arid Region of the United States, with a More Detailed Account of the Lands of Utah*. In this remarkable document, he debunked every fantasy about the desert West, roundly proclaiming that agricultural methods used elsewhere would not work in the West. He compiled a rainfall map, which delineated areas with enough rainfall to allow nonirrigated agriculture. In other areas west of the one-hundredth meridian, he noted, irrigation agriculture was practicable, but only on a small scale. He recommended forming local irrigation districts to administer water management, with no farmer having more than eighty acres (thirty-two hectares). In giving his recommendations, Powell made use of Mormon experience, where everyone farmed on family properties, authority over water being a communal concern. The Mormons had shown the sustainable way to farm the West.

A firestorm of criticism burst over Powell's head from politicians and land speculators, who believed that humanity should have complete mastery over the earth. But his science was undeniable. There was insufficient water to irrigate most of the West. Thus, argued his opponents, a network of dams and reservoirs would be needed to collect it and use it most effectively. At an irrigation conference in Los Angeles in 1893, Powell delivered a seminal address, calling on simple calculations to make his point: "When all the rivers are used, when all the creeks in the ravines, when all the brooks, when all the springs are used, when all the reservoirs along the streams are used, when all the canyon waters are taken up, when all the artesian waters are taken up, when all the wells are sunk or dug that can be dug in all this arid region, there is still not sufficient water to irrigate all this arid region."[7] (Artesian wells are those drilled through impermeable strata into levels that receive water from a higher altitude, so there is pressure to force the water to flow upward.)

Powell was shouted down. His enemies in Congress forced him to resign from the Geological Survey. But history has shown that he was right. He had studied subsistence agriculture in the West and knew well how easily sustainability falters. Later, Powell wrote prophetically, "All the waters of all the arid lands will eventually be taken from their natural channels."[8] In his day, damming the Colorado River would have been a wild idea, for the concrete arch dam had not been invented.

With Powell silenced, the speculators turned to gravity, led by water-hungry Los Angeles. Ambitious officials cast their eyes on Owens Lake, a huge, turquoise body of water flanked by the Sierra Nevada and the White Mountains. A pioneer settler, Beveridge R. Spear, wrote, "The lake was alive with wild fowl. Ducks were by the square mile, millions of them. When they rose in flight, the roar of their wings . . . could be heard ten miles away."[9] Owens Lake was the shrunken remnant of a much larger lake from the late Ice Age. Only a trickle of water entered the lake, which was more saline than fresh, but it supported a species of brine shrimp and flies that attracted millions of waterfowl. The desert surroundings of the lake had little attraction except for hunters who preyed on the myriad birds, but the river was another matter. By 1899, local Anglo colonists grew alfalfa, hay, and fruit on more than 40,000 acres (16,187 hectares) of irrigated land, much of their produce going to a large silver-mining camp in Tonopah, Nevada.

Some 250 miles (402 kilometers) away lay Los Angeles, a mere village, but a location with seemingly endless potential—if it could be supplied with water. Mormon farmers had irrigated the dry but fertile soils of the Los Angeles Basin with rainfall from the San Bernardino Mountains, where artesian wells flowed strongly. By 1900, the flow of the Los Angeles River was dropping fast, thanks to uncontrolled pumping of groundwater. There was no other water nearby, for deserts surrounded the city on three sides, the Pacific hemming it in on the fourth. To divert water from the Colorado and Kern rivers was an impossibility at the time, for the technology to pump water over mountain ranges did not yet exist. There remained the Owens River, where there was enough water for a million people. In 1904, William Mulholland, a self-taught hydraulic engineer, became head of the newly formed Los Angeles Department of Water and Power. In the years that followed, Los Angeles used every possible form of legal and illegal chicanery to divert water from the Owens River in the face of powerful opposition. In 1907, work began on an aqueduct that would span 223 miles (359 kilometers) of some of the most challenging landscape in the American West with tunnels and siphons. Up to six thousand men labored on the aqueduct, opened by Mulholland on November 5, 1913. Before a crowd of forty thousand, he turned to the mayor of Los Angeles when the water appeared on the sluiceway and said, simply, "There it is. Take it."[10]

By 1925, Los Angeles, with its 1.2 million inhabitants, was growing fifteen times faster than Denver, eleven times faster than New York. The surrounding county enjoyed the highest agricultural output in the nation. All of this agriculture depended on irrigation water flowing from a river over 200 miles (321 kilometers) away, the Los Angeles aqueduct being, in many senses, the downstream extension of the Owens River. The aqueduct came online during a series of wet years, so at first there was plenty of water to go around. When droughts returned during the 1920s, the Owens Valley farmers suffered at once. As Los Angeles grew, the river water went entirely to the south; then the Department of Water and Power sank wells and started draining the Owens Valley aquifer. The last Owens River ranchers quit during the 1950s; promiscuous aquifer pumping continued. After 1913, the lake became a huge salt flat. Now huge clouds of alkaline dust blow through the desiccated valley.

The Owens River turned Los Angeles into a megalopolis, located in an arid landscape where, by the rules of common sense, no city should ever stand. Los Angeles hefts enough political clout to capture any river within 600 miles (966 kilometers). Today, the city receives water not only from the Owens River but also via aqueducts from the Colorado River and from the California Aqueduct, which runs from the Sacramento– San Joaquin Delta to Lake Perris, in Riverside County, 444 miles (715 kilometers) to the south.

Dams were inevitable as agribusiness expanded and the technology to build them with concrete emerged early in the twentieth century. A period of aggressive federal reclamation and dam projects began with the passage of the Newlands Act in 1902. A boom in irrigation agriculture was soon under way. Congress set up a Colorado River Commission twenty years later. Its members hammered out the Colorado River Compact, which allocated water between the seven states that shared the Colorado River Basin. This landmark agreement transformed parts of the desert West into a bustling hub of industrial-scale farming and urban growth, culminating in the building of the Hoover Dam between 1931 and 1936.[11] Other dams followed. The era of industrial water management was truly under way, for the benefit not of small farmers but of large agribusinesses. In the interests of reaching agreement, the Colorado River Compact left many difficult water-allocation questions unanswered, notably those affecting Indian tribes and Mexico. A paper mountain range, the Law of the River, now comprises at least fifty major governing documents, to say nothing of the complex environmental regulations and Clean Water Acts enacted since the 1960s. Small armies of lawyers and scholars now devote their careers to Western water law, as shrinking supplies muddy already intricate legal waters. The compact is an anachronism, developed by commissioners who lived in a sparsely inhabited, predominantly rural West. They would be startled by today's Phoenix and the endlessly proliferating subdivisions of suburban Los Angeles and Salt Lake City. In short order, however, they would hear about rapid growth, the paving of deserts, and rapidly shrinking water supplies. They would also learn about disturbing climatic changes—intensifying droughts and higher temperatures. All of them had experienced dry cycles, but, for some reason, they never factored them into their calculations.

As expansion accelerated and pumping technology improved, both cities and farmers turned to finite aquifers to make up shortfalls from mountain runoff and other sources. Aquifer pumping has intensified exponentially. Wells penetrate deeper and deeper. The groundwater that supplies much of Tucson's water used to lie 492 feet (150 meters) below the surface. Now it's at 1,476 feet (450 meters) and falling. As a result, Arizona began regulating groundwater overdrafts as long ago as 1980. Thanks to conservation measures including low-flow appliances, Tucson has a water consumption of about 99 gallons (375 liters) per person per day, one of the lowest in the United States. But the city continues to grow, and the demand is still too great. In California's San Joaquin Valley, pumping has already outgrown natural replenishment by more than half a trillion gallons a year, and counting—and that's before you factor in projections that the West is entering a period much drier than recent centuries.

The Colorado River's past offers a sobering portrait of the future. By chance, the Colorado River Commission forged the compact during the longest high-flow period of the past four and a half centuries. We know this because of Colorado Plateau tree-ring records, which reveal very severe dry periods between 1564 and 1600 and from 1868 to 1892. Between 1844 and 1848, the Colorado's flow was only two thirds of its long-term average. Further back in time, there was a prolonged and severe dry cycle between 1139 and 1154. Tree-stump rings in Owens Lake show a major drought from before 910 to 1100 and another long, intensely dry spell from before 1250 to about 1350. Until the last century, the West's population was small enough to swing with the punches. By 2050, millions more people will be competing for Colorado water. Global warming will have reduced the flow of the already drought-depleted river by a further 10 percent.[12] Those two banes of irrigation agriculture, silt load and salinization, will have intensified. No question, our grandchildren and great-grandchildren will live in a very different hydrological world. Quite apart from renegotiating the now-obsolete Colorado River Compact, we will have to break the habits of our lifetimes and use water very differently. For instance, geologist James Lawrence Powell (no relation to John Wesley) estimates that reducing agricultural allocations and the amount of city water going to landscaping from 50 percent to 5 percent would save nearly 20 percent of the annual flow of the Colorado River alone.[13]

GROUNDWATER IS ALSO vanishing further to the east, from Colorado and New Mexico to Texas, Oklahoma, and Nebraska, where the vast Ogallala aquifer, under the Great Plains, supports hundreds of communities, as well as large cities and major agricultural and mining activities. The Ogallala supplies about a third of the nation's groundwater used for irrigation. Eighty-two percent of the people living within the boundary of the aquifer obtain their drinking water from it. The water-permeated sedimentary rocks of the Ogallala Formation are up to one thousand feet (three hundred meters) thick and cover an area of about 174,000 square miles (450,000 square kilometers). U.S. Geological Survey experts have calculated that irrigation alone sucked about 21 million acre-feet (260 cubic kilometers) of water from the Ogallala in 2000, a figure slightly larger than the historic annual discharge rate of the Colorado River. Some hydrologists believe that the aquifer will dry up in about twenty-five years.[14]

The Ogallala lies under some of the most productive ranching and cereal-cultivation land in the United States, which could not be cultivated until sophisticated centrifugal pumps that brought up thousands of gallons of water a minute came along between the 1930s and 1950s, at the same time that rural electrification brought power to remote farms. Landscapes once ravaged during the Dust Bowl became some of the most agriculturally productive regions in the world. By 1980, there were more than 17,300,000 acres (7 million hectares) of irrigated land. At first, both the farmers and the government assumed that the waters of the aquifer were infinite. They soon found out they were wrong, for the water level fell as much as 5 feet (1.5 meters) a year during periods of maximum extraction. In some parts of northern Texas, the water has already dried up, while in others artesian wells sink deeper and deeper. The aquifer has shrunk by about 9 percent since 1950. Widespread conservation methods such as more efficient drip irrigation and reducing acreage have slowed depletion, but the fact remains that the Ogallala is a finite resource that will eventually run dry or, at the most, provide much-diminished water supplies at increasingly high cost. When that happens, where will the water that supports hundreds of thousands of people come from? These questions are as much political and social as they are economic, for the interests of many stakeholders are involved, among them, for example, communities on the shores of the Great Lakes, an obvious potential water source.

Even without sustained warming and mounting aridity, much of the United States lives on borrowed time, for climatic changes are already raising the hydrological stakes. And all of this also raises another disturbing question: How much longer will the United States be able to provide surplus food to other hungry parts of the world as it has done in the past? In the foreseeable future, the water supplies that feed crops will potentially be more valuable than the thirsty cotton and rice they irrigate.

THIS MAY SEEM like a surprising statement, but the fact is that the world's supply of freshwater is finite. As global population rises, the demand for food, and for the water that produces it, grows inexorably. Globally, farming accounts for 70 percent of our withdrawals from this fixed "bank account," this in the face of ever-greater domestic and industrial usage. Water tables are falling in many parts of the world, among them China, India, and the United States. The Himalayan glaciers will shrink massively in the next century, reducing natural water storage in the mountains. The shortfalls will have to come from groundwater and surface storage. Many great rivers have drastically diminished flows. Bangladesh is suffering from the diversion of Ganges River water and increased salinization. Underground aquifers in many places are shrinking so rapidly that NASA satellites are detecting changes in the earth's gravity. The Water Resources Group has estimated that India may face a 50 percent lag in water availability relative to demand by 2030 and that global availability may lag demand by as much as 40 percent—the statistics have been questioned. Sixty years ago, the world's population was about 1.25 billion people; few people, even in arid lands, worried about water supplies. Then came the Green Revolution, with its new, high-yielding crops, which depend on fertilizers and a great deal more irrigated farming. Global populations skyrocketed—to nearly seven billion by 2009, with a projected nine billion by 2050. By the same year, the five hundred million people living in areas chronically short of water in the year 2000 will have grown by 45 percent to four billion. A billion of us currently go hungry because there is not enough water to grow food. Much of the world's water is still unpriced, but it is now the most valuable commodity in the world.

To compound the problem, 60 percent of the world's population lives in crowded river basins shared by several countries, often with daggers drawn at one another. The Tigris, Euphrates, and Nile rivers nurtured the world's first civilizations with waters that came from distant sources.[15] These societies, with their smaller numbers, certainly suffered from water shortages, which caused desert nomads to encroach on settled lands and city authorities to ration grain. Today, millions of people depend on these rivers, most of them living far downstream of their sources in other countries. Turkey has ambitious plans for the Tigris and the Euphrates that involve numerous dams for both irrigation and hydroelectric uses. Both Syria and Iraq will receive much less water as a result, the former about 40 percent less. With a rapidly growing population, arid landscapes, seriously depleted groundwater, and less water coming from the Euphrates, Syria faces curtailed agricultural development and possible food shortages. Iraq has a larger population and lies even further downstream, while depending almost entirely on river water. As much as 80 percent of its Euphrates flow may be lost and, eventually, much of that of the Tigris. Tensions over water allocations between Turkey, Syria, and Iraq are high and have yet to be resolved by international agreement, a difficult task when the countries downstream claim that the Euphrates and the Tigris are international waters—90 percent of Euphrates water comes from Turkey.

Egypt depends entirely on the Nile for its irrigation agriculture, as it has for more than five thousand years. The ancient Egyptians produced one crop a year from the inundation. However, by maintaining a high river level and using pumps powered by fossil fuels, today's Egypt can grow two or three crops a year to feed its exploding population. The Egyptian government built the Aswan High Dam by the First Cataract in the 1970s, creating the waters of Lake Nasser. If there is sufficient water in the lake, the authorities can irrigate crops year-round. However, what does not flow downstream is the rich, fertilizing silt that always arrived with the inundation. Now Egypt's soils are depleted; sardine fisheries off the Mediterranean coast are in trouble because the now-absent silt fed the nutritious plankton growing just offshore; the delta coast is eroding. Furthermore, with a virtually constant river level, soils are waterlogged and the salt level is rising, owing to poor drainage. The rising groundwater level is causing salt-bearing moisture to seep into ancient Egypt's temples.

The Aswan Dam has not solved Egypt's water problems, which are compounded by global warming and more frequent droughts. The country is now even more dependent on water from afar. In 1978, Egypt and Sudan began construction on the Jonglei Canal, a huge project to divert water from tributaries of the White Nile past the Sudd swamps far upstream, but the project foundered with the outbreak of civil war in southern Sudan. There is still hope of its completion, but that lies far in the future. Most of Egypt's water actually comes from the Blue Nile, which rises in Ethiopia. As in countries downstream, the upper Blue Nile catchment suffers from rapid population growth, as well as environmental degradation and severe soil erosion. Many people along the river are starving. One solution may be irrigation, but any large-scale dam and storage schemes would have a serious impact on water supplies downstream. Both Egypt and Sudan are nervous, although one could argue that storing water far upstream in environments where there is less surface evaporation might be beneficial for all parties involved. The long-term solution lies in close cooperative relationships between all the countries lying in or around the entire Nile Basin. Given the chronic instability and political unrest in the region, such an agreement may be a long time in coming, but it offers the only logical solution to allocation problems. An equally likely outcome is water wars, triggered by significant curtailments of water supplies flowing downstream—and these would surely solve nothing.

The problems are just as acute elsewhere, especially in arid areas with growing populations where boreholes and aquifers are thought to be the answer. Seemingly a miraculous solution, but not if the drawdown exceeds the replenishment rate, as is the case with the groundwater beneath a now-sinking Mexico City's twenty million inhabitants, and with Bangkok, Buenos Aires, and Jakarta, where pollution and rising salt levels combine with overdrafting. In China, deep groundwater levels have dropped as much as 295 feet (90 meters) in places. We have perforated the earth's surface with boreholes to deplete a resource that we all, ultimately, hold in common.

Now we stand at the threshold of what I have called a third stage in our relationship with water, one where, apparently, cataclysm looms on every side. Vivid doomsday scenarios espoused by numerous writers have Phoenix imploding as its water supplies fail, the Nile drying up, tens of thousands of people crossing national boundaries to find water. Futurist after futurist warns that water wars are a certainty in coming centuries. Alas, at least some of these cataclysms could descend upon us if we persist in denying the seriousness of the water crisis and deluding ourselves into thinking that uncontrolled growth and more dams are the solution. They are not. Yes, there will be shortfalls, people will go thirsty and die, but in the end, as has happened so many times in the past, human ingenuity, quite apart from technology, will find solutions. And in the process, we will develop new, much more respectful relationships with water, even if they do not necessarily have the profound spiritual intertwinings of earlier times.

In the short term, there are four potential ways of improving the situation, but none of them will solve the problem of chronic overdrawing.[16] One lies in spending large sums on systematic improvements to storage and delivery, to the infrastructure behind water supplies. Underground reservoirs have potential. So do simple things like replacing leaking pipes, lining earth-bottomed canals, and irrigating plants at their roots with just the right amount of water, among many others. A second solution also makes sense: Make farming less thirsty by using drought-resistant, higher-yielding, even genetically modified crops. This is much easier said than done, for significant technological breakthroughs lie a long way in the future. Also, we should not forget that planting more crops means more use of water, since each plant transpires vapor into the atmosphere through photosynthesis. One possible solution may lie in developing plants that can grow using saline water, but, again, this development is in the future.

Then there's another seemingly attractive option: desalinization. Surprisingly, this has been around a long time. None other than Aristotle remarked that "salt water, when it turns to steam, becomes sweet, and the steam does not form salt water when it condenses."[17] Julius Caesar's legions drank freshwater condensed from seawater during his siege of Alexandria in 48–47 B.C.E. As self-appointed visionaries keep reminding us, desalinization seems like the answer to all our problems, but, in spite of improvements in efficiency, there remain significant environmental and technical problems. Desalinization, which involves creating and recondensing steam, consumes prodigious amounts of energy, even in its most efficient iterations, so it is currently confined to nations where oil is cheap and abundant. Nearly half the existing desalinization plants are in the Arabian Peninsula and along the Persian Gulf, especially in Saudi Arabia and the Gulf States. In most other places, the cost of desalinization is three or four times that of conventional water sources. The cost of oil is rising, so the alternatives are either coal or nuclear power, both of which have their own environmental consequences and political baggage. Desalinization plants operate along seacoasts; many of the most water-hungry areas are far inland, thereby adding huge transport costs to the already high price of a gallon of desalinized water. What, also, are we to do with the brine resulting from desalinization, which has to be disposed of? Once again, breakthroughs lie in the long-term future. At present, desalinization is no panacea, for it contributes only about 0.4 percent of global water supplies.

Finally, there's conservation, which involves both profound changes in our mind-sets and completely new attitudes toward water as a marketable commodity. Water is scarce, but it is also a complicated thing to market. It is difficult to move, hard to measure accurately in large quantities, and complex to price and charge for. Most people resent paying for water, as they think it should be free or very cheap. Even in dry parts of the world where every drop is precious, the price of water seldom reflects its true scarcity. However, we are entering an era of potentially ferocious trading in water rights and a time when water could cost more than oil, as managing demand becomes an international priority. It's no coincidence that privately owned companies are quietly and aggressively purchasing water rights in many countries. Increasingly, municipal and other authorities are pricing water according to usage. Judging from experience in Australia, Los Angeles, and other water markets, the strategy leads to reduced water use, especially when combined with measures to save water such as reduced-flow toilets and strict timetables for watering. Like oil, water is a commodity that will be the subject of market forces, with price mechanisms that will bring supply and demand into balance. Once water is properly priced, the economics of international trade may encourage water-rich countries to produce water-intensive goods and arid ones to make those that are water light.

Mind-sets are notoriously difficult to change, especially in societies accustomed to abundance and seemingly unlimited water supplies. Using the forces of the marketplace and stricter allocations will not be strategies of first choice, especially in urban settings with high levels of poverty. Nor will conservation in the form of another commonly proposed measure, yet more dam construction, prove effective. History from the near and remote past tells us that dams are no panacea, for they silt up, and silt has to be removed or the dam becomes shallower and ever less useful. And, even more important, where is the water to fill them going to come from? No dam ever creates water; it merely captures what is a finite supply. How can new dams provide more water in the era of prolonged global drought that lies ahead? Besides, there's adequate dam capacity in the American West to store any water that will come from the smaller snowpacks of future decades. Short of creating more water, more efficient allocation, extensive water recycling for landscaping and other purposes, drastic reductions in agribusiness water subsidies, and miserly use of current supplies are some viable strategies for the future. And this kind of conservation, on scales small and large, is the responsibility of us all. Our survival depends upon it.

We have much to learn about water conservation from the experience of our ancestors. This book is about the traditional and the simple, about ingenious solutions that supplied drinking water, irrigated fields, and disposed of wastewater. Humans have managed water successfully for thousands of years in ways that are often far off the historical radar screen. We learn from their experiences that it is the simple and ingenious that often works best—local water schemes, decisions about sharing and management made by kin, family, and small communities. These experiences also teach us that self-sustainability is attainable.

Such ingenuity comes in many forms. It may be a simple idea in the field or, in this day and age, more likely the inspiration for a social and political initiative that changes the way people think. We are moving into an entirely new water future, where equity of use, sustainability to protect future generations, and affordability for everyone are major components. A new paradigm for water management, based on well-defined priorities in which all stakeholders have a voice, will have to govern our future water use.[18] Our salvation lies in long-term thinking, in decisive political leadership, and in a reordering of financial priorities, for, after all, investing heavily in water management will alleviate much disease and poverty automatically. Above all, the future will need a massive shift in our relationship with water to one that equates, at least approximately, with that of those who went before us—characterized by a studied caring and reverence. Only one thing is certain: Descartes was wrong. We will never master the earth.

Acknowledgments

My first and most fundamental debt is to my good friend Vernon Scarborough, Mayanist extraordinaire, who first introduced me to water history through his writings. He has inspired and encouraged me throughout the gestation of this book.

Elixir also originated in a series of lectures about the Medieval Warm Period that I delivered to Mesa Community College, in Arizona; to the faculty of Columbia River College, in Pasco, Washington; and to the California Water Policy Conference. Without the stimulating discussions and questions that arose from the audiences at these talks, this book would never have happened. This is a synthesis of my personal experiences with water in the past, a surprisingly little-known subject that is very much my own interpretation of a maze of academic literature from disciplines ranging from archaeology and history to paleoclimatology, hydrology, and beyond. I am, of course, responsible for the accuracy of these pages and the conclusions therein, and will, as always, hear in short order from those kind, often anonymous individuals who delight in pointing out errors large and small. Let me thank them in advance.

The research for this book involved consulting dozens of specialists, who were invariably helpful and courteous, as well as being remarkably efficient in replying to e-mails. I am deeply grateful for your assistance and for you taking me seriously, even when my questions were startlingly elementary. It's impossible to name everyone, but my long list of indebtedness includes Richard Cummins, Robin Coningham, Carole Crumley, Eve Darien-Smith, Nadia Durrani, Rick Effland, Ronald Fletcher, Jonathan Kenoyer, Sherleen Lerner, Lisa Lucero, George Michaels, Paul Sinclair, Sesh Velamoor, Tony Wilkinson, and Kenneth and Ruth Wright. I know I have omitted some folks—please forgive me. The participants at a symposium, "Water: The Crisis Ahead," at the Foundation for the Future, in Bellevue, Washington, in April 2010, provided valuable insights.

As always, Shelly Lowenkopf was at my side, arguing, spotting inconsistencies, and serving as an inspiring literary coach. We have worked on many books together; this was probably the hardest. Steve Brown drew the maps and drawings with his customary skill. My agent, Susan Rabiner, supported this book from the moment when it was merely a couple of pages of narrative. My debt to her is enormous.

Peter Ginna and Pete Beatty encouraged me from the beginning and provided editorial guidance at the most critical stages. Their merciless and perceptive criticism has made the book immeasurably better. I value my association with them more than I can say.

As always, Lesley and Ana suffer through deadlines and chapters that don't work, as does our ever-changing family of beasts. I appreciate their tolerance and understanding, even if they make me sponsor rabbit festivals.

Brian Fagan
Santa Barbara, California

Notes

This book travels through both well-trodden and profoundly obscure byways of the remote and not-so-remote past, using a combination of firsthand observation, wide-ranging discussions with specialists, and a huge, often very specialized literature. The references that appear below provide a cross section of the literature as of mid-2010 and contain useful bibliographies for those wishing to probe deeper.

General Works

There are numerous general works on water and its history. The following are a few recent titles, in addition to those cited in specific chapters below.

Maude Barlow, *Blue Covenant: The Global Water Crisis and the Coming Battle for the Right to Water* (New York: New Press, 2009).

Marq de Villiers, *Water: The Fate of Our Most Precious Resource* (Boston: Houghton Mifflin, 2001).

Alice Outwater, *Water: A Natural History* (New York: Basic Books, 1996).

Fred Pearce, *When the Rivers Run Dry: The Defining Crisis of the Twenty-first Century* (Boston: Beacon Press, 2007).

E. C. Pielou, *Fresh Water* (Chicago: University of Chicago Press, 1998).

Norman Smith, *Man and Water: A History of Hydro-technology* (London: Peter Davies, 1975).

Steven Solomon, *Water: The Epic Struggle for Wealth, Power, and Civilization* (New York: Harper, 2010).

Veronica Strang, *The Meaning of Water* (New York: Berg, 2004).

James G. Workman, *Heart of Dryness: How the Last Bushmen Can Help Us Endure the Coming Age of Permanent Drought* (New York: Walker and Co., 2009).

Epigraphs

1. Geoffrey Chaucer, *The Canterbury Tales*, "The Miller's Tale," lines 405–13, online edition, http:/www.librarius.com/cantran.mttrfs.htm.

2. Arlette Ottino, *The Universe Within: A Balinese Village Through Its Ritual Practices* (Paris: Karthala, 2000), 14.

Preface

1. Holy Quran 47:15.

2. Rachel Carson, *Silent Spring* (Boston: Houghton Mifflin, 1962), 39.

3. Lao-tzu, *Tao Te Ching*, trans. Stephen Mitchell, 1995, chap. 78, 1–6, online edition, http://academic.brooklyn.cuny.edu/core9/phalsall/texts.taote-v3 .html.

Chapter 1: The Elixir of Life

1. Brian Fagan, D. W. Phillipson, and S. G. H. Daniels, *Iron Age Cultures in Zambia*, vol. 2, *Dambwa, Ingombe Ilede, and the Tonga* (London: Chatto and Windus, 1969).

2. Thayer Scudder, *The Ecology of the Gwembe Tonga* (Manchester, UK: Manchester University Press, 1962), is a classic account of these people and their environment. Biblical quote from *Ecclesiastes* I, 1:7.

3. Daniel Hillel, *The Rivers of Eden* (New York: Oxford University Press, 1994), chap. 1, describes the realities of water admirably.

4. Karl Wittfogel, *Oriental Despotism: A Comparative Study of Total Power* (New Haven, CT: Yale University Press, 1957).

5. Some selected references on Australian Aborigines and water:

David Bruno, *Landscapes, Rock-Art and the Dreaming: An Archaeology of Preunderstanding* (London: Leicester University Press, 2002).

Philip Clark, *Where the Ancestors Walked: Australia as an Aboriginal Landscape* (Crows Nest, Australia: Allen and Unwin, 2003).

Richard A. Gould, "Subsistence Behavior Among the Western Desert Aborigines of Australia," *Oceania* 39, no. 4 (1969): 253–57.

———, *Yiwara: Foragers of the Australian Desert* (New York: Charles Scribner's Sons, 1969).

Marcia Langton, "Earth, Wind, Fire, Water: The Social and Spiritual Construction of Water in Aboriginal Societies," in *The Social Archaeology of Australian*

Indigenous Societies, ed. Bruno David et al. (Canberra, Australia: Aboriginal Studies Press, 2006), 139–60.

Robert Tonkinson, *The Mardudjara Aborigines: Living the Dream in Australia's Desert* (New York: Holt, Rinehart and Winston, 1978).

6. R. Berndt, "The Sacred Site: The Western Arnhem Land Example," *Australian Aboriginal Studies* 29 (1970): 15.

7. Quotes in this paragraph from http://www.derm.qld.gov.au/waterwise /resources/pdf/activities/p3australianwaterstories.pdf.

8. M. Barceló et al., *The Design of Irrigation Systems in al-Andalus* (Barcelona: Universitat Autonoma de Barcelona, 1998).

9. G. W. W. Barker and G. D. B. Jones, "The UNESCO Libyan Valleys Survey IV," *Libyan Studies* 14 (1983): 39–68.

10. Rev. 22:1

11. Psalm 68:9–10.

Chapter 2: Farmers and Furrows

1. Graeme Barker, *The Agricultural Revolution in Prehistory* (New York: Oxford University Press, 2006), chap. 4. See also Brian Fagan, *The Long Summer* (New York: Basic Books, 2004), chap. 5.

2. Bill Finlayson and Steven Mithen, eds., *The Early Prehistory of Wadi Faynan, Southern Jordan: Excavations at the pre-Pottery Neolithic A Site of WF16 and Archaeological Survey of Wadis Faynan, Ghuwayr and Al Bustan* (Oxford, UK: Oxbow Books, 2007).

3. Finlayson and Mithen, *Early Prehistory*, Part 1.

4. For example, Audrey Richards, *Land, Labor and Diet in Northern Rhodesia* (Oxford, UK: Oxford University Press, 1939).

5. M. Widgren, "Towards a Historical Geography of Intensive Farming in Eastern Africa," in *Islands of Intensive Agriculture in Eastern Africa*, ed. M. Widgren and J. E. G. Sutton (Oxford, UK: James Currey, 2004), 11.

6. A large literature surrounds Engaruka. Two excellent summaries: J. E. G. Sutton, "Engaruka: The Success and Abandonment of an Integrated Irrigation System in an Arid Part of the Rift Valley," in *Islands*, Widgren and Sutton, 114–32; Sutton, "Engaruka: Farming by Irrigation in Maasailand," in *The Archaeology of Drylands: Living at the Margin*, ed. Graeme Barker and David Gilbertson (London: Routledge, 2000), 201–19.

Chapter 3: "Whoever Has a Channel Has a Wife"

1. Joseph Thomson, *Through Masai Land* (London: Sampson Low, 1885), 310. Thomson (1858–95) was a Scottish geologist turned explorer. He led a successful expedition from Dar es Salaam to Lakes Nyasa and Tanganyika in 1878 and a second one to the northern shores of Lake Victoria. It was on this expedition that he encountered irrigation canals. Thomson was an efficient explorer who abhorred violence.

2. William M. Adams and David M. Anderson, "Irrigation Before Development: Indigenous and Induced Change in Agricultural Water Management in East Africa," *African Affairs* 87, no. 349 (1998): 519–35.

3. B. E. Kipkorir, R. C. Soper, and J. W. Ssenyonga, eds., *Kerio Valley: Past, Present and Future* (Nairobi: Institute of African Studies, University of Nairobi, 1983).

4. Quotes from Wilhelm Östberg, "The Expansion of Marakwet Hill-Furrow Irrigation in the Kerio Valley of Kenya," in *Islands of Intensive Agriculture in Eastern Africa*, ed. M. Widgren and J. E. G. Sutton (Oxford: James Currey, 2004), 36.

5. Ibid., 47.

6. Robert F. Gray, *The Sonjo of Tanganyika: An Anthropological Study of an Irrigation-Based Society* (Oxford, UK: Oxford University Press, 1963).

7. Ibid., 110.

8. Matthew Davies, "The Irrigation System of the Pokot, Northwestern Kenya," *Azania* 43 (2008): 50–76.

Chapter 4: Hohokam: "Something That Is All Gone"

1. Frank Hamilton Cushing (1857–1900) was an anthropologist at the Smithsonian Institution. He lived among the inhabitants of Zuni Pueblo from 1879 to 1884, becoming anthropology's first participant observer. His much-reprinted book *My Adventures in Zuni* (Palo Alto, CA: American West, 1970) is a classic of early anthropology. See also *Zuñi: Selected Writings of Frank Hamilton Cushing*, ed. Jesse Green (Lincoln: University of Nebraska Press, 1979), 12–14.

2. The esoterica of history: Many early army officers, and even Theodore Roosevelt, believed that the town was named after noted officer and writer John Derby, who wrote under the nom de plume John Phoenix. The issue is unresolved.

3. This chapter draws heavily on Suzanne K. Fish and Paul R. Fish, eds., *The Hohokam Millennium* (Santa Fe, NM: School for Advanced Research Press, 2007).

4. Omar Turney's map appears in Fish and Fish, *Hohokam*, 2–3.

5. Henry D. Wallace, "Hohokam Beginnings," in *Hohokam*, Fish and Fish, 13–22.

6. Ibid., 14.

7. Las Capas: Jonathan B. Mabry, ed., *Las Capas: Early Irrigation and Sedentism in a Southwestern Floodplain* (Tucson, AZ: Center for Desert Archaeology, 2008). Valencia Vieja: Henry D. Wallace, *Roots of Sedentism: Archaeological Excavations at Valencia Vieja, a Founding Village in the Tucson Basin of Southern Arizona* (Tucson, AZ: Center for Desert Archaeology, 2003).

8. Daniel Lopez, "Huhugam," in *Hohokam*, Fish and Fish, 120–21.

9. Douglas B. Craig and T. Kathleen Henderson, "Houses, Households, and Household Organization," in *Hohokam*, Fish and Fish, 31–37.

10. Paul R. Fish and Suzanne K. Fish, "Community, Territory, and Polity," in *Hohokam*, Fish and Fish, 38–47.

11. Discussion in Craig and Henderson, "Houses," 30ff.

12. David E. Doyel, "Irrigation, Production, and Power in Phoenix Basin Hohokam Society," in *Hohokam*, Fish and Fish, 83–91.

13. David R. Wilcox, "The Mesoamerican Ballgame in the American Southwest," in *The Mesoamerican Ballgame*, ed. Vernon Scarborough and David R. Wilcox (Tucson: University of Arizona Press, 1991), 101–25.

14. Emil W. Haury, *The Hohokam: Desert Farmers and Craftsmen* (Tucson: University of Arizona Press, 1976), is a fundamental source on the Snaketown ball court.

15. Mark D. Elson, *Expanding the View of Hohokam Platform Mounds: An Ethnographic Perspective* (Tucson: University of Arizona Press, 1998). See also the same author's "Into the Earth and Up into the Sky: Hohokam Ritual Architecture," in *Hohokam*, Fish and Fish, 49–56.

16. Stephanie M. Whittlesey, "Hohokam Ceramics, Hohokam Beliefs," in *Hohokam*, Fish and Fish, 65–74.

17. Ibid., 72.

Chapter 5: The Power of the Waters

1. Chapter 5 draws heavily on J. Stephen Lansing's definitive monograph, *Priests and Programmers: Technologies of Power in the Engineered Landscape of Bali*, rev. ed. (Princeton, NJ: Princeton University Press, 2007). See also Lansing's *The Balinese* (New York: Harcourt Brace, 1995) for a more general account. Rice agriculture: Lucien M. Hanks, *Rice and Man: Agricultural Ecology in Southeast Asia* (Chicago: Aldine, 1972).

2. Margaret Mead, "The Arts in Bali," *Yale Review* 30, no. 2 (1940): 335.

3. Lansing, *Priests*, 59.

4. These paragraphs are based on Lansing, *Priests*, chap. 3. Quote from 73.

5. Ibid., 76.

6. Ibid., 93.

7. From the chronicle of the Paseks of the Black Wood. Ibid., 86–87.

8. Clifford Geertz, *Negara: The Balinese Theater State in the Nineteenth Century* (Princeton, NJ: Princeton University Press, 1980), 83.

9. The description of irrigation systems and their history is based on Lansing, *Priests*, chap. 3.

10. Geertz, *Negara*, 125.

11. Lansing, *Priests*, chap. 3.

12. Ibid., 70.

13. Ibid., pp. 67–70.

Chapter 6: Landscapes of Enlil

1. Composite translation, *Dispute Between Winter and Summer* (Oxford, UK: Electronic Text Corpus of Sumerian Literature, 2000), lines 190–225. http://www-etcsl.orient.ox.ac.uk/section5/tr533.htm.

2. This chapter owes much to T. J. Wilkinson, *Archaeological Landscapes of the Near East* (Tucson: University of Arizona Press, 2003), which sets the stage for a new generation of landscape archaeology in the region. For Near Eastern water supplies, see also Hillel, *Rivers*, chap. 2 (see chap. 1, n. 3).

3. J. N. Postgate, *Early Mesopotamia: Society and Economy at the Dawn of History* (London: Routledge, 1992), 11.

4. J. A. Black et al., *Enlil in the E-kur (Enlil A)* (Oxford, UK: Electronic Text Corpus of Sumerian Literature, 1998), paragraph 12.

5. Austen Henry Layard, *Nineveh and Babylon* (London: John Murray, 1853), 479.

6. This section: Wilkinson, *Archaeological Landscapes*, chap. 5.

7. Postgate, *Early Mesopotamia*, 183.

8. Modern-day observations confirm that gradient is all-important. Gravity irrigation is hard to maintain when slopes exceed 2 degrees (3.5 percent) and very difficult indeed when the gradient exceeds 4 degrees (about 7 percent). James A. Neely, "Sassanian and Early Islamic Water-Control and Irrigation Systems on the Deh Luran Plain, Iran," in *The Organization of Power: Aspects of Bureaucracy in the Ancient Near East*, ed. McGuire Gibson and Robert D. Biggs (Chicago: Oriental Institute Studies in Ancient Oriental Civilization, 1987), 21–43.

9. Descriptions in Wilkinson, *Archaeological Landscapes*, chap. 5.

10. David Oates and Joan Oates, "Early Irrigation Agriculture in Mesopotamia," in *Problems in Economic and Social Archaeology*, ed. G. de G. Sieveking et al. (London: Duckworth, 1976), 109–52.

11. Martha Prickett, "Settlement During the Early Periods," in *Excavations at Tepe Yahya, Iran, 1967–1975: The Early Periods*, ed. Thomas W. Beale (Cambridge, MA: American Society for Prehistoric Research, 1986), 215–46.

12. Wilkinson, *Archaeological Landscapes*, chap. 2.

Chapter 7: The Lands of Enki

1. Samuel Kramer, *The Sumerians* (Chicago: University of Chicago Press, 1963), 240.

2. Once again, this chapter draws significantly on multiple chapters of Wilkinson, *Archaeological Landscapes* (see chap. 6, n. 2).

3. Seton Lloyd, *Mounds of the Near East* (Edinburgh: Edinburgh University Press, 1963).

4. Climatic shifts summarized in Fagan, *Long Summer*, chap. 7 (see chap. 2, n. 1), and Wilkinson, *Archaeological Landscapes*, chap. 2.

5. Robert Adams's surveys of Mesopotamian landscapes are classics. See Robert McC. Adams, *Land Behind Baghdad: A History of Settlement on the Diyala Plains* (Chicago: University of Chicago Press, 1965), and *Heartland of Cities: Surveys of Ancient Settlement and Land Use on the Central Floodplain of the Euphrates* (Chicago: University of Chicago Press, 1981).

6. Harriett Crawford, *Sumer and the Sumerians*, 2nd ed. (Cambridge: Cambridge University Press, 2004), surveys the literature. See also Robert McC. Adams and Hans J. Nissen, *The Uruk Countryside: The Natural Setting of Urban Societies* (Chicago: University of Chicago Press, 1972).

7. Kramer, *Sumerians*, 106–7.

8. Characteristics of fields: Wilkinson, *Archaeological Landscapes*, 52ff.

9. Hans J. Nissen, *The Early History of the Ancient Near East, 9000–2000 B.C.* (Chicago: University of Chicago Press, 1988).

10. Mario Liverani, *Uruk: The First City*, trans. Zainab Bahrani and Marc Van De Mieroop (London: Equinox, 2006).

11. Kramer, *Sumerians*, 152.

12. Nissen, *Early History*, 141.

13. Piotr Steinkeller, "New light on the Hydrology and Topography of Southern Babylonia in the Third Millennium," *Zeitschrift fur Assyriologie* 91 (2001): 22–84. Quote from 14.

14. J. S. Cooper, "Reconstructing History from Ancient Inscriptions: The Lagash-Umma Border Conflict," *Sources and Monographs on the Ancient Near East* 2, n. 1 (1983): 47–54.

15. Ibid., 50.

16. These sections are based on Harvey Weiss and Marie Agnès Courty, "The Genesis and Collapse of the Akkadian Empire: The Accidental refraction of Historical Law," in *Akkad: The First World Empire*, ed. Mario Liverani (Padua, Italy: Sargon, 1993), 131–56. See also Harvey Weiss et al., "The Genesis and Collapse of Third Millennium North Mesopotamian Civilization," *Science* 261, no. 5124 (1993): 995–1004.

Chapter 8: "I Caused a Canal to Be Cut"

1. Herodotus, *The Histories*, trans. Robin Waterfield (Oxford, UK: Oxford University Press, 1998), bk. 2, line 111, 136.

2. Barry Kemp, *Ancient Egypt: The Anatomy of a Civilization* (London: Routledge, 2006), offers a definitive account of ancient Egypt for our purposes.

3. J. Donald Hughes, "Sustainable Agriculture in Ancient Egypt," *Agricultural History* 66, no. 2 (1992): 13.

4. A. H. Gardiner, *Hieratic Papers in the British Museum*, 3rd Series, *Chester Beatty Gift* (London: Egypt Exploration Society, 1935), 41.

5. David Oates, *Studies in the Ancient History of Northern Iraq* (London: British Academy, 1968). See also Gunther Garbrecht, "The Water System at Tuspa (Urartu)," *World Archaeology* 11, no. 3 (1980): 306–12; Jason Ur, "Sennacherib's Northern Assyrian Canals: New Insights from Satellite Imagery and Aerial Photography," *Iraq* 67, no. 1 (2005): 317–45.

6. Ibid., 46.

7. Thorkild Jacobsen and Seton Lloyd, *Sennacherib's Aqueduct at Jerwan* (Chicago: University of Chicago Press, 1935). Quote from 20.

8. A useful survey of *qanats* used for this section: Peter Beaumont et al., *Qanat, Kariz and Khattara: Traditional Water Systems in the Middle East and North Africa* (London: Middle East Center, School of Oriental and African Studies, 1989). See also Hillel, *Rivers*, 192–93 (see chap. 1, n. 3).

9. C. E. Stewart, *Through Persia in Disguise, with Reminiscences of the Indian Mutiny* (London: Routledge, 1911). Quote from 341.

10. Jørgen Laessøe, "The Irrigation System at Ulhu, 8th Century B.C.," *Journal of Cuneiform Studies* 5 (1951): 25–32. Quotes from 27.

11. Ibid., 29.

12. Sassanians: Peter Christiansen, *The Decline of Iranshahr* (Copenhagen: Museum Tusculum Press, 1993); Touraj Daryaee, *Sasanian Persia: The Rise and Fall of an Empire* (London: I. B. Tauris, 2009).

13. For the Diyala irrigation works, see the surveys described in Adams, *Land* (see chap. 7, n. 5).

14. James A. Neely, "Sassanian and Early Islamic Water-Control and Irrigation Systems on the Deh Luran Plain," in *Irrigation's Impact on Society*, ed. Theodore Downing and McGuire Gibson (Tucson: University of Arizona Press, 1974, 21–42). See also James A. Neely and H. T. Wright, *Early Settlement and Irrigation on the Deh Luran Plain: Village and State Societies in Southwestern Iran* (Ann Arbor: University of Michigan Museum of Anthropology, 1994).

Chapter 9: The Waters of Zeus

1. Homer, *Odyssey*, trans. Robert Fagles (New York: Viking, 1996), bk. 7, lines 147–56, 183.

2. This chapter relies heavily on Dora P. Crouch's *Water Management in Ancient Greek Cities* (New York: Oxford University Press, 1993), a definitive work by any standards. This section is based on chap. 3.

3. Ibid. 22.

4. Joan Evans, *Time and Chance* (London: Longmans Green, 1943), is a comprehensive biography of this remarkable archaeologist. For his Balkan travels, see A. J. Evans, *Through Bosnia and the Herzegóvina on Foot During the Insurrection* (London: Longmans Green, 1876). Quote from 72.

5. Sir Arthur Evans, *The Palace of Minos at Knossos*, vol. 1, *The Neolithic and Early and Middle Minoan Ages* (Oxford: Clarendon Press, 1921), 143.

6. Ibid., 228.

7. A. H. Angelakis, Y. M. Savvakis, and G. Charalampskis, "Aqueducts During the Minoan Era," *Water Science and Technology: Water Supply* 7, no. 1 (2007): 95–101.

8. Herodotus, *Histories*, bk. 3, line 60, 194 (see chap. 8, n. 1).

9. Vitruvius, *De Architectura*, trans. Frank Granger (Cambridge, MA: Harvard University Press, 1970), bk. 8, chap. 2, 67. The karst discussion that follows is based on Crouch, *Water Management*, chap. 7.

10. This section draws on Crouch, *Water Management*, chaps. 5 and 18.

11. Plutarch, *Greek Lives: A Selection of Nine Greek Lives*, trans. Robin Waterfield (New York: Oxford University Press, 1998), 23.

12. Corinth and Syracuse: Crouch, *Water Management*, chap. 8.

13. A good source on water-lifting devices is Örjan Wikander, ed., *Handbook of Ancient Water Technology* (Boston: Brill, 2000).

14. D. R. Shackleton Bailor, ed., *Anthologia Latina* (Stuttgart, Germany: B. G. Teubneri, 1982), 224.

15. W. W. Westermann, "Land Reclamation in the Fayum Under Ptolemies Philadelphus and Euergetes I," *Classical Philology* 12, no. 4 (1917): 426–30.

16. Julius Firmicus Maternus was a Christian, Latin writer and noted astronomer of the fourth century C.E. His *De Errore Profanum Religionum*, trans. Clarence A. Forbes (New York: Newman Press, 1970), appeared in about 346 C.E. Quote from bk. 4, chap. 13, line 3.

17. Crouch, *Water Management*, 113–14.

Chapter 10: Aquae Romae

1. Pliny the Younger, *The Letters of Pliny the Younger*, vol. 5, chap. 5, "Letter to Domitus Apolinaris," http://ancienthistory.about.com/library/bl/bl_text_plinyltrs1.htm.

2. Sextus Julius Frontinus, *De Aqueducta Urbis Romae*, trans. C. E. Bennett and Mary B. McElwain (Cambridge, MA: Harvard University Press, 1925), bk. 1, 16.

3. This passage is based on A. Trevor Hodge, *Roman Aqueducts and Water Supply* (London: Duckworth, 2002). This book is a definitive account of Roman aqueducts, which I have used extensively for this chapter. See also Charles E. Bennett, *The Stratagems and the Aqueducts of Rome* (Cambridge, MA: Harvard University Press, 1969).

4. Rome's water supply is studied in detail by Christer Bruun, *The Water Supply of Ancient Rome* (Helsinki: Societas Scientiarum Fennica, 1991). I have drawn extensively on his work here.

5. Discussion in Hodge, *Roman Aqueducts*, chap. 5.

6. Vitruvius, *De Architectura*, bk. 8, chap. 4, line 1 (see chap 9., n. 9).

7. Plumbers and pipes: Bruun, *Water Supply*, 304–68.

8. Vitruvius, *De Architectura*, bk. 8, chap. 4, line 1 (see chap 9., n. 9).

9. Rodolfo Lanciani, *The Ruins & Excavations of Ancient Rome* (London: Macmillan, 1897).

10. *Castella*: Hodge, *Roman Aqueducts*, chap. 10.

11. Frontinus, *De Aqueducta Urbis Romae*, bk. 2, 87.

12. Discussion in Hodge, *Roman Aqueducts*, 305.

13. Garrett G. Fagan, *Bathing in Public in the Roman World* (Ann Arbor: University of Michigan Press, 1999), is the standard work on Roman baths. Lu-

cius or Marcus Annaeus Seneca (circa 3 B.C.E.–65 C.E.) was a Roman dramatist and courtier, famous for his tragedies. Quote from John W. Basore, trans. *Epistles* vol. 1, *Epistles 1–65* (Cambridge, MA: Harvard University Press, 1917), Epistle 56, lines 1–2.

14. Barry Cunliffe, *Roman Bath* (London: Society of Antiquaries, 1969).

15. Hodge, *Roman Aqueducts*, chap. 9.

16. Ibid., chap. 12.

Chapter 11: *Waters That Purify*

1. Quote from Gregory Possehl, *The Indus Civilization: A Contemporary Perspective* (Walnut Creek, MD: AltaMira Press, 2002), 6–7.

2. Ralph T. H. Griffith, trans, *The Hymns of the RigVeda*, rev. ed. (New Delhi: Motilal Banarsidass, 1973). Quote from Hymn 7–49, line 2. I am grateful to Professor T. N. Naranimhan for bringing this reference to my attention.

3. Ibid., Hymn 2–13, line 12.

4. Quote from Possehl, *Indus*, 8. See also the same author's "The Transformation of the Indus Civilization," *Journal of World Prehistory* 11, no. 4 (1997): 425–72.

5. Possehl, *Indus*, offers an excellent description. Another, more popular account: Jane McIntosh, *A Peaceful Realm: The Rise and Fall of the Indus Civilization* (Boulder, CO: Westview Press, 2002). Irrigation: U. Alam, P. Sahota, and P. Jeffrey, "Irrigation in the Indus Basin: A History of Unsustainability?" *Water Science and Technology: Water Supply* 7, no. 1 (2007): 211–18.

6. Kathleen D. Morrison, "Supplying the City: The Role of Reservoirs in an Indian Urban Landscape," *Asian Perspectives* 32, no. 2 (2006): 133–51.

7. Rita P. Wright et al., "Water Supply and History: Harappa and the Beas Regional Survey," *Antiquity* 82, no. 1 (2008): 37–48.

8. Possehl, *Indus*, chap. 11, has an excellent description.

9. M. Jansen, "Water Supply and Sewage Disposal at Mohenjo-Daro," *World Archaeology* 21, no. 2 (1989): 177–92.

10. This section is based on Ravindra Singh Bisht, "The Water Structures and Engineering of the Harappans at Dholavira (India)," *South Asian Archaeology* 1 (2005): 11–26.

11. This section is based on Kathleen D. Morrison, "Making Places and Making States: Agriculture, Metallurgy, and the Wealth of Nature in South India," in *The Wealth of Nature: How Natural Resources Have Shaped Asian History, 1600–2000*, ed. P. Boomgaard and G. Bankoff (New York: Palgrave Macmillan),

81–99. See also C. M. Sinapoli and K. D. Morrison, *The Vijayanagara Metropolitan Survey*, vol. 1 (Ann Arbor: University of Michigan Museum of Anthropology, 2007).

12. A general reference: E. R. Leach, "Hydraulic Society in Ceylon," *Past and Present* 15, no 1 (1959): 2–26. Recent research: Robin Coningham, *Anuradhapura: The British–Sri Lankan Excavations at Anuradhapura Salgaha Watta 2*, vol. 1, *The Site* (Oxford: British Archaeological Reports International Series 824, 1999). A summary of the region appears in R. A. E. Coningham, "South Asia: From Early Villages to Buddhism," in *The Human Past*, ed. C. J. Scarre (London and New York: Thames and Hudson, 2005), 518–51.

13. R. A. E. Coningham et al., "The State of Theocracy: Defining an Early Medieval Hinterland in Sri Lanka," *Antiquity* 81, no. 4 (2007): 699–719. See also R. L. Brohier, *Ancient Irrigation Works in Ceylon* (Colombo: Government Publications Bureau, 1934).

14. A basic description of the site appears in Jayasinghe Balasooriya, *The Glory of Ancient Polonnaruwa* (Polonnaruva: Sooriya Printers, 2004).

15. Two excellent descriptions: Charles Higham, *The Civilization of Angkor* (Berkeley: University of California Press, 2001); Michael D. Coe, *Angkor and the Khmer Civilization* (London and New York: Thames and Hudson, 2003).

16. Roland Fletcher et al., "The Water Management Network of Angkor, Cambodia," *Antiquity* 82, no. 4 (2008): 658–70.

17. Edward R. Cook et al., "Asian Monsoon Failure and Megadrought During the Last Millennium," *Science* 328, no. 5977 (2010): 486–89.

Chapter 12: China's Sorrow

1. Biographical details of the Chinese rulers mentioned in this chapter come from Ann Paludin, *Chronicle of the Chinese Emperors* (London and New York: Thames and Hudson, 1998).

2. Mark Elvin, introduction, in *Sediments of Time: Environment and History in Chinese Society*, ed. Mark Elvin and Liu Ts'ui-jung (Cambridge: Cambridge University Press, 1998), 1–30. Quote from 9.

3. Charles Greer, *Water Management in the Yellow River Basin of China* (Austin: University of Texas Press, 1979).

4. Tristram R. Reilly, *Geoarchaeology at Sanyangzhuang: A Preliminary Report* (St. Louis, MO: Washington University, Department of Anthropology, 2010). I am grateful to Professor Reilly for drawing this important site to my attention.

5. Greer, *Water Management*, chap. 2, summarizes the major early developments.

6. Dujiangyan irrigation system on the Min River: *China Heritage Newsletter* 1 (Canberra: Australian National University, 2005).

7. Grand Canal: Joseph Needham, with Wang Ling and Lu Gwei-Djen, *Science and Civilization in China*, vol. 4, *Physics and Physical Technologies*, pt. 3, *Civil Engineering and Nautics* (Cambridge: Cambridge University Press, 1971), 307–10. See also Mark Elvin, *The Pattern of the Chinese Past* (Stanford, CA: Stanford University Press, 1973), 54–55.

8. Elvin, Pattern, 55.

9. This description of Qing waterworks near Beijing is based on Lillian M. Li, *Fighting Famine in North China* (Stanford, CA: Stanford University Press, 2007).

10. My account of Chen Hongmou is based on the definitive biography: William T. Rowe, *Saving the World: Chen Hongmou and Elite Consciousness in Eighteenth Century China* (Stanford, CA: Stanford University Press, 2001).

11. Pierre-Etienne Will, *Bureaucracy and Famine in Eighteenth-Century China*, trans. Elborg Forster (Stanford, CA: Stanford University Press, 1990).

12. Technology: Joseph Needham, *Science and Civilization in China*, vol. 4, *Physics and Physical Technologies*, pt. 2, *Mechanical Engineering* (Cambridge: Cambridge University Press, 1965). Water-raising machinery is described in chap. 27 (g and h), 330–407.

13. Peter C. Perdue, *Exhausting the Earth: State and Peasant in Hunan, 1500–1850* (Cambridge, MA: Harvard University Council on East Asian Studies, 1987).

14. Ibid., 221.

15. Ibid., 231.

16. Mike Davis, *Late Victorian Holocausts: El Niño Famines and the Making of the Third World* (New York: Verso, 2001).

17. Quotes in this paragraph from Davis, *Late Victorian*, 72.

Chapter 13: The Water Lily Lords

1. Bernal Díaz, *The Conquest of New Spain*, trans. J. M. Cohen (Baltimore: Pelican Books, 1963). Quotes in this paragraph from 214.

2. Lisa J. Lucero and Barbara W. Fash, eds., *Precolumbian Water Management* (Tucson: University of Arizona Press, 2006). See also Richard Townsend, *The Aztecs* (London and New York: Thames and Hudson, 2009).

3.   Vernon Scarborough, *Flow of Power: Ancient Water Systems and Landscapes* (Santa Fe, NM: School of American Research Press, 2003), was a fundamental source for this chapter.

4.   General accounts of the Maya civilization abound. Michael D. Coe, *The Maya*, 7th ed. (London and New York: Thames and Hudson, 2005), offers an excellent summary.

5.   Lisa J. Lucero, *Water and Ritual: The Rise and Fall of Classic Maya Rulers* (Austin: University of Texas Press, 2006). See also her "The Political and Sacred Power of Water in Classic Maya Society," in *Precolumbian*, Lucero and Fash, 116–28.

6.   William A. Saturno, "The Dawn of Maya Gods and Kings," *National Geographic* 209, no. 1 (2001): 68–77. See also William A. Saturno, Karl Taube, and David Stuart, *The Murals of San Bartolo, Guatemala*, pt. 1, *The North Wall* (Barnardsville, NC: Center for Ancient American Studies, 2005).

7.   Nicholas P. Dunning et al., "Environmental Variability Among Bajos in the Southern Maya Lowlands and Its Implications for Ancient Maya Civilization and Archaeology," in *Precolumbian*, Lucero and Fash, 81–99. See also N. S. Dunning et al., "Arising from the Bajos: Anthropogenic Change of Wetlands and the Rise of Maya Civilization," *Annals of the Association of American Geographers* 92, no. 2 (2002): 267–83.

8.   Vernon L. Scarborough, *Archaeology at Cerros, Belize, Central America*, vol. 3, *The Settlement System in a Late Preclassic Maya Community* (Dallas: Southern Methodist University Press, 1991).

9.   Edzná and El Mirador: Scarborough, *Flow*, pp. 50ff.

10.   Ibid., 50ff and 110ff. See also Peter Harrison, "Aspects of Water Management in the Southern Maya Lowlands," in *Precolumbian*, Lucero and Fash, 71–122.

11.   Scarborough, *Flow*, 110ff.

12.   Lucero, *Water and Ritual*, has an extended discussion.

13.   Barbara W. Fash and Karla L. Davis-Salazar, "Copán Water Ritual and Management," in *Precolumbian*, Lucero and Fash, 129–43.

14.   Kirk D. French et al., "Archaeological and Epigraphic Evidence for Water Management and Ritual at Palenque," in *Precolumbian*, Lucero and Fash, 144–52.

15.   The Maya collapse is the subject of an enormous literature. The arguments in this section are covered in my *The Great Warming* (New York: Bloomsbury Press, 2008), chap. 8, where references will be found.

16.   James A. Webster et al., "Stalagmite Evidence from Belize Indicating Significant Droughts at the Time of Preclassic Abandonment, the Maya Hiatus, and the Classic Maya Collapse," *Palaeogeography, Palaeoclimatology, Palaeoecology*

250, nos. 1–4 (2007): 1–17. See also Heather Pringle, "A New Look at the Mayas' End," *Science* 324, no. 5926 (2004): 545–46; Holley Moyes, "Charcoal as a Proxy for Use-Intensity in Ancient Maya Cave Ritual," in *Religion, Archaeology, and the Material World*, ed. Lara Fogelin (Carbondale, IL: Center for Archaeological Investigations, 2008), 139–58.

17. Deforestation: Robert J. Oglesby et al., "Collapse of the Maya: Could Deforestation Have Contributed?" *Journal of Geophysical Research* 115 (2010), D 12106, doi: 10. 1029/2009 JD011942.

Chapter 14: Triumphs of Gravity

1. Clark L. Erickson, "Applied Archaeology and Rural Development: Archaeology's Potential Contribution to the Future," *Journal of the Steward Anthropological Society* 20, nos. 1–2 (1992): 1–16. See also William M. Deneven, Kent Mathewson, and Gregory Knapp, eds., *Prehispanic Agricultural Fields in the Andean Region* (Oxford, UK: British Archaeological Reports, International Series 359, 1987); Clark L. Erickson, "Prehistoric Landscape Management in the Andean Highlands: Raised Field Agriculture and Its Environment Impact," *Population and Environment* 13, no. 4 (1992): 285–300.

2. Clark L. Erickson, "Archaeological Methods for the Study of Ancient Landscapes of the Llanos de Mojos in the Bolivian Amazon," in *Archaeology in the Lowland Amazon Tropics*, ed. Peter W. Stahl (Cambridge: Cambridge University Press, 1995), 66–95.

3. Richard L. Burger, *Chavín and the Origins of Andean Civilization* (London and New York: Thames and Hudson, 1992). See also William J. Conklin and Jeffrey Quilter, *Chavín: Art, Architecture, and Culture* (Los Angeles: Cotsen Institute of Archaeology, UCLA, 2008).

4. Michael Moseley, *The Inca and Their Ancestors*, rev. ed. (London and New York: Thames and Hudson, 2001), is an authoritative summary.

5. The best account is still Walter Alva and Chris Donnan, *Royal Tombs of Sipán* (Los Angeles: Fowler Museum of Cultural History, UCLA, 1989), although there have been other spectacular discoveries since then. See also Steve Bourget and Kimberly L. Jones, eds., *The Art and Architecture of the Moche* (Austin: University of Texas Press, 2008).

6. Michael Moseley and Kent C. Day, eds., *Chan Chan: Andean Desert City* (Albuquerque: University of New Mexico Press, 1982); Moseley, *Inca*, chap. 9. See also my *Great Warming*, chap. 9 (see chap. 13, n. 15).

7. My Nasca discussion is based on Katharina Schreiber and Josué Lancho Rojas, *The Puquios of Nasca* (Lanham, MA: Lexington Books, 2003).

8. Kevin J. Vaughn, *The Ancient Andean Village: Marcaya in Prehispanic Nasca* (Tucson: University of Arizona Press, 2009).

9. This section is based on Kenneth R. Wright, *Tipon: Water Engineering Masterpiece of the Inca Empire* (Reston, VA: ASCE Press, 2006).

10. Machu Picchu's hydrology: Kenneth R. Wright and Alfredo Valencia Zagarra, *Machu Picchu: A Civil Engineering Marvel* (Reston, VA: ASCE Press, 2000), upon which this account is based.

11. Ibid., 36.

12. Ibid., 38.

Chapter 15: The Waters of Islam

1. Holy Quran. Quotes in this paragraph are from Suras 11:9, 30:48, and 47 (see preface, n. 1).

2. Hadith are narratives stemming from the words and deeds of the Prophet. Most were gathered into large collections during the eighth and ninth centuries.

3. General histories of Islam: David Levering Lewis, *God's Crucible: Islam and the Making of Europe, 570–1215* (New York: Norton, 2008); Jonathan Lyons, *The House of Wisdom: How the Arabs Transformed Western Civilization* (New York: Bloomsbury Press, 2009).

4. Lewis, *God's Crucible*, 133.

5. Tony Wilkinson, "From Highland to Desert: The Organization of Landscape and Irrigation in Southern Arabia," in *Agricultural Strategies*, ed. Joyce Marcus and Charles Stanish (Los Angeles: Cotsen Institute of Archaeology, UCLA, 2006), 38–68. See also I. Hehmeyer, E. J. Keall, and D. Rahimi, "Ghayl Ba Wazir: Applied Qanat Technology in the Fissured Karst Landscape of Southern Yemen," *Proceedings of the Seminar for Arabian Studies* 32 (2002): 83–97.

6. Tony Wilkinson, J. C. Edens, and M. Gibson, "The Archaeology of the Yemen High Plains: A Preliminary Chronology," *Arabian Archaeology and Epigraphy* 8 (1997): 99–142.

7. Andrew M. Watson, *Agricultural Innovation in the Early Islamic World: The Diffusion of Crops and Farming Techniques, 700–1130* (Cambridge: Cambridge University Press, 1983), 89.

8. Ibid. 113.

9. Lewis, *God's Crucible*, 205.

10. M. Barceló et al., *The Design of Irrigation Systems in al-Andalus* (Barcelona: Universitat Autonoma de Barcelona, 1998). See also Thomas F. Glick and Helena Kirchner, "Hydraulic Systems and Technologies of Islamic Spain: His-

tory and Archaeology," in *Working with Water in Medieval Europe*, ed. Paolo Squatriti (Leiden, Netherlands: Brill, 2000), 267–330.

11. D. Fairchild Ruggles, "Waterwheels and Garden Gizmos: Technology and Illusion in Islamic Gardens," in *Wind & Water in the Middle Ages*, ed. Steven A. Walton (Tempe: Arizona Center for Medieval and Renaissance Studies, 2006), 69–88. See also Thomas F. Glick, *Irrigation and Society in Medieval Valencia* (Cambridge, MA: Belknap Press, 1970).

Chapter 16: "Lifting Power . . . More Certain than That of a Hundred Men"

1. Quotes in this paragraph are from Walton, introduction in *Wind & Water*, xviii (see chap. 15, n. 11).

2. I am grateful to Professor William Jordan of Princeton University for researching and translating this quote for me. Martin Bourquet et al., eds., *Recuil des Historians des Gaules et de la France*, vol. 21 (Paris: Imprimarie Royale, 1738–1904), 725.

3. This passage is based on Roberta J. Magnussen, *Water Technology in the Middle Ages* (Baltimore: Johns Hopkins University Press, 2001). Quote from 19.

4. Wolfgang Braunfels, *Monasteries of Western Europe: Regime and Architecture, 900–1900*, trans. Alastair Laing (London and New York: Thames and Hudson, 1972). Quote from 18.

5. Sylvia Landsberg, *The Medieval Garden* (London and New York: Thames and Hudson, 1996).

6. Brian Fagan, *Fish on Friday* (New York: Basic Books, 2006), covers medieval fish farming and religious observances associated with meatless days.

7. Terry S. Reynolds, *Stronger than a Hundred Men* (Baltimore: Johns Hopkins University Press, 1983), is the definitive source on early pumping technology. Quote from 5.

8. Walton, *Wind & Water*, 9.

9. Ibid., 26.

10. Matthew 26:26.

11. Reynolds, *Stronger*, 5.

12. Reynolds, *Stronger*, chaps. 2 and 3, covers this ground.

13. Ibid., chap. 3.

14. Ibid.

15. Roberta J. Magnussen, "Public and Private Urban Hydrology: Water Management in Medieval London," in *Wind & Water*, Walton, 171–88.

16. Newcomen engines, simpler predecessors of James Watt's engines, were widely used to pump out mines as early as 1715.

17. Steven Johnson, *The Ghost Map* (New York: Riverhead Trade, 2007), offers a lively account of the cholera epidemics and their consequences.

18. Phyllis Deane, *The First Industrial Revolution*, 2nd ed. (Cambridge: Cambridge University Press, 1979).

19. This section is based on Reynolds, *Stronger*, chap. 5. For steam technology, see William Rosen, *The Most Powerful Idea in the World: A Story of Steam, Industry, and Invention* (New York: Random House, 2010).

20. Theodore Steinburg, *Down to Earth: Nature's Role in American History*, 2nd ed. (New York: Oxford University Press, 2008), covers this issue admirably.

Chapter 17: Mastery?

1. Ezekiel 36:25–26.

2. Oliver Goldsmith, *History of the Earth, and Animated Nature* (London: F. Wingrave, 1794), 143.

3. David Blackbourn, *The Conquest of Nature* (New York: W. W. Norton, 2006).

4. Ibid., 97.

5. This section is based on Marc Reisner, *Cadillac Desert: The American West and Its Disappearing Water* (New York: Viking, 1986). Despite a torrent of popular books on water in the West, this is still the best account.

6. These sections are based on James Lawrence Powell, *Dead Pool: Lake Powell, Global Warming, and the Future of Water in the West* (Berkeley: University of California Press, 2008).

7. Ibid., 46.

8. Ibid., 49.

9. Reisner, *Cadillac Desert*, 61.

10. Ibid., 89.

11. Michael Hiltzik, *Colossus: Hoover Dam and the Making of the American Century* (New York: Free Press, 2010), is the latest account of the construction.

12. This figure is based on the Colorado River Open Source Simulator (CROSS), described in Powell, *Dead Pool*, 184. Estimate on 231.

13. Ibid., 242.

14. The High Plains Aquifer Information Network is a valuable information source on the current status of the aquifer. http://www.kgs.ku.edu/HighPlains /hiplain.

15. Hillel, *Rivers*, 242 (see chap. 1, n. 3).

16. This section is based on John Grimond, "For the Want of a Drink," *Economist*, May 20, 2010, http://www.economist.com/node/16136302.

17. Hillel, *Rivers*, 251.

18. A summary of many of the issues surrounding water today can be found in the Executive Summary of "Water: The Crisis Ahead," a workshop organized by the Foundation for the Future, Bellevue, Washington, April 21–23, 2010, http://www.futurefoundation.org/documents/HUM_WaterWorkshop_ExecutiveSummary.pdf.

Index

Note: Page numbers in *italics* indicate an illustration, photograph, or map.

A Note on the Author

Brian Fagan is emeritus professor of anthropology at the University of California–Santa Barbara. Born in England, he did fieldwork in Africa and has written about North American and world archaeology and many other topics. His books on the interaction of climate and human society have established him as the leading authority on the subject; he lectures frequently around the world. He is the editor of *The Oxford Companion to Archaeology* and the author of *Cro-Magnon*; *The Great Warming*; *Fish on Friday: Feasting, Fasting, and the Discovery of the New World*; *The Little Age*; and *The Long Summer*, among many other titles.